Computer Monographs

GENERAL EDITOR: Stanley Gill, M.A., Ph.D.
ASSOCIATE EDITOR: J. J. Florentin, Ph.D., Birkbeck College, London

Data Transmission

Data Transmission

M.D. Bacon
and
G.M. Bull
Department of Computer Science
The Hatfield Polytechnic

Macdonald · London and
American Elsevier Inc. · New York

© M. D. Bacon and G. M. Bull 1973

Sole distributors for the British Isles and Commonwealth
Macdonald & Co. (Publishers) Ltd.
49–50 Poland Street, London W.1

Sole distributors for the United States and Dependencies
American Elsevier Publishing Company Inc.
52 Vanderbilt Avenue, New York, N.Y. 10017

All remaining areas
Elsevier Publishing Company
P.O. Box 211, Jan van Galenstraat 335, Amsterdam, The Netherlands

Macdonald ISBN 0 356 04387 8
American Elsevier ISBN 0 444 19564 5
Library of Congress Catalog Card No 72–124 18

All Rights Reserved. No part of this publication may be reproduced, stored in a retrieval system, or transmitted, in any form or by any means, electronic, mechanical, photocopying, recording or otherwise, without prior permission of the publishers.

Made and printed in Great Britain by
Balding+Mansell, London and Wisbech

Contents

Preface

1 Concepts in communications   1
1.1 Introduction   1
1.2 Waves—concepts of frequency, wavelength and period   1
1.3 Fourier analysis   5
1.4 Information and channel capacity   12
    1.4.1 Information   12
    1.4.2 Channel capacity   15
1.5 Transmission media   17

2 Communication codes   18
2.1 The Baudot code   18
2.2 The ASCII code   18
2.3 Binary and binary coded decimal   20
2.4 The EBCDIC code   21
2.5 Control characters   22
2.6 Other codes   24

3 Modulation   25
3.1 Introduction   25
3.2 Amplitude modulation   26
3.3 Frequency modulation   31
3.4 Phase modulation   34

4 Pulse techniques   36
4.1 Introduction   36
4.2 Pulse amplitude modulation   37
4.3 Pulse width modulation   40
4.4 Pulse position modulation   40
4.5 Pulse code modulation   41
4.6 Modulation with orthogonal digital carriers   44

5 Noise, error detection and correction — 45
- 5.1 Noise — 45
- 5.2 Error detection and correction — 49

6 Parallel and serial transmission — 57
- 6.1 Definitions — 57
- 6.2 Parallel transmission — 57
- 6.3 Serial transmission—asynchronous — 59
- 6.4 Serial transmission—synchronous — 60
- 6.5 Modems — 62

7 Store-and-forward and circuit switched networks — 65
- 7.1 Introduction — 65
- 7.2 Circuit switching — 65
- 7.3 The store-and-forward concept — 66

8 Multiplexing and multiplexers — 68
- 8.1 Multiplexing — 68
- 8.2 Multiplexers — 69
- 8.3 Line sharing — 75

9 Asynchronous line interfaces — 79
- 9.1 Hardware interface — 79
- 9.2 DEC PDP8 — 88
- 9.3 Main processor software — 95

10 Synchronous line interfaces — 98
- 10.1 Hardware interface — 98
- 10.2 Software — 103

11 Analog and parallel interfaces — 108
- 11.1 Introduction — 108
- 11.2 Analog-to-digital conversion — 108
- 11.3 Digital-to-analog conversion — 109
- 11.4 Software — 111
- 11.5 Parallel data transmission — 113

12 Telecommunication organisations 117
12.1 International Telecommunication Union 117
12.2 National organisations 118
12.3 International services 119
12.4 Survey of facilities 119

References 125

Glossary of Terms 128

Bibliography 132

Index 134

Preface

This book is intended for both undergraduate and postgraduate Computer Science students. It is based on lectures given to students following the B.Sc. and the M.Sc. courses in Computer Science at The Hatfield Polytechnic.

Our experience is that Computer Science students have a very vague understanding of the meaning of frequency, wavelength, amplitude and phase; for this reason we have included the topics in Chapter One. Frequency analysis and the dual relation between the frequency domain and the time domain is so important to the communication engineer that it is also included. Likewise the topics information (together with Shannon's formula), and codes. Later chapters discuss some of the standard communication techniques—modulation, pulse techniques, error correction and also some of the communication terms such as serial asynchronous transmission and multiplexing. We then apply the communication concepts to data transmission systems such as multiplexers and interfaces. The last chapter discusses data transmission facilities.

In a book of this size each topic is necessarily brief. Further reading matter is given at the end of the book.

The book is also suitable for engineering students who are concerned with telecommunications, and for those professional people who are concerned with computer communications.

We wish to thank Digital Equipment Company Ltd and the British Post Office for their generous permission to quote from their publications.

We should also like to thank Dr R. W. Sharp for reading the draft and making valuable suggestions and comments.

G. M. B.
M. D. B.
Feb 1973

1 Concepts in communications

1.1 Introduction

Human beings have always communicated by speech and signals. Writing enabled messages to be recorded and carried to their destination by men, pigeons, etc. More recently the messages have been transmitted along wires and also broadcast by wireless devices. Machines have been developed which are operated by messages, so that now there are three forms of message interaction, man/man, man/machine and machine/machine. The passing of a message is called communicating and data transmission is a form of communication. The word data is used here for any sequence of symbols, whatever the meaning. They may be speech patterns suitably coded, or the complicated sequence used by a computer as its program of operations.

For a communication process to take place there are three basic elements, a transmitter (or message source), a message medium and a receiver (or message acceptor). A message passed from transmitter to receiver in the form of a signal is always affected in some way by its environment. The signal can be examined by special devices at intermediate points of its path, and will be found to be in general a very complex structure. A signal can also be represented mathematically.

1.2 Waves—concept of frequency, wavelength and period

The word waves usually brings to mind those formed by the sea. Water waves are just one example of many different forms of wave motion; they are, in fact, very complicated owing to the action of many different forces. Complicated waves can be thought of as a combination of many simple waves.

The most important feature of waves is that they are a sequence or

DATA TRANSMISSION

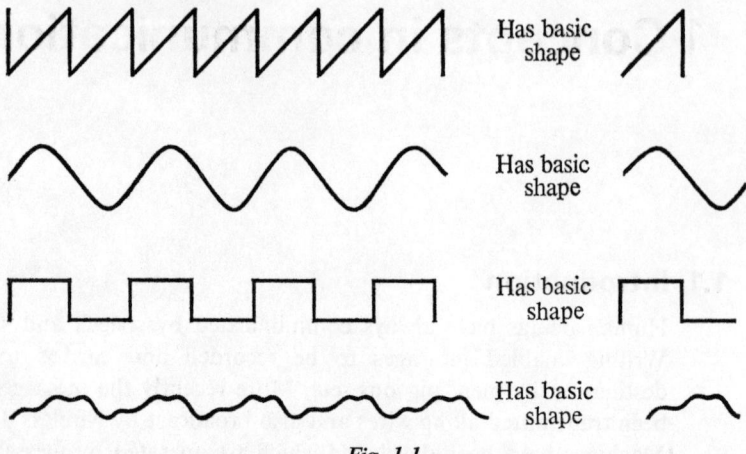

Fig. 1.1

repetition of some basic shape, they are said to be periodic. (See Fig. 1.1.)

The length of one basic shape unit is called the *wavelength* in whatever units the system is operating; for example seconds, or feet.

An important simple wave can be obtained by considering a point P rotating round a circular path (see Fig. 1.2). Such a wave occurs commonly in, for example, the domestic electricity supply, because it is produced by rotating a conductor in a magnetic field.

The projection of P on to the x axis gives the length

$$OX = OP \cos \theta.$$

The projection of P on to the y axis gives the length

$$OY = OP \sin \theta.$$

Let the rotation time start at $t = 0$ when P is at D.

If P rotates with a constant angular velocity ω radians per second then in time t seconds the point will have moved through an angle $\theta = \omega t$ radians; $a \cos \omega t$ plotted as a function of time is shown in Fig. 1.3.

The *wavelength* in this case is $2\pi/\omega$ secs; a is called the *amplitude*. The time taken for one basic shape to appear is called the *period*; it is, in this simple case, the time taken for P to travel one complete circle; that is for θ to go from zero to 2π, or t from 0 to $2\pi/\omega$. The period is often written as T.

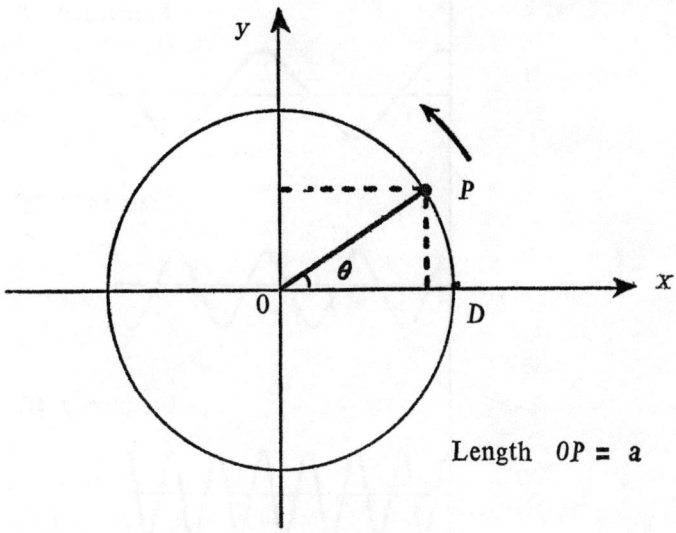

Fig. 1.2

The number of times the simple basic shape repeats itself per second is called the *frequency*. If one basic shape is $T = 2\pi/\omega$ secs, then the number of such shapes per second is $1/T = \omega/2\pi = f$. (See Fig. 1.4.) ω is often called the angular frequency since it has the dimensions of frequency, the units being radians per second; it is related to the true frequency by the equation $\omega = 2\pi f$.

Therefore associated with cosine or sine shaped curves are the features wavelength, amplitude, period and frequency. Such waves are called Simple Harmonic because there is only one frequency. It is

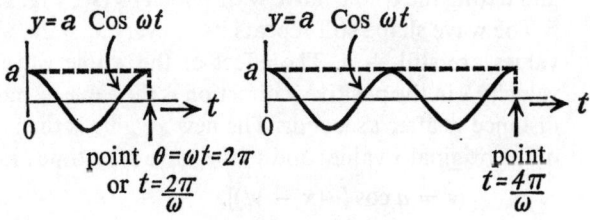

Fig. 1.3

DATA TRANSMISSION

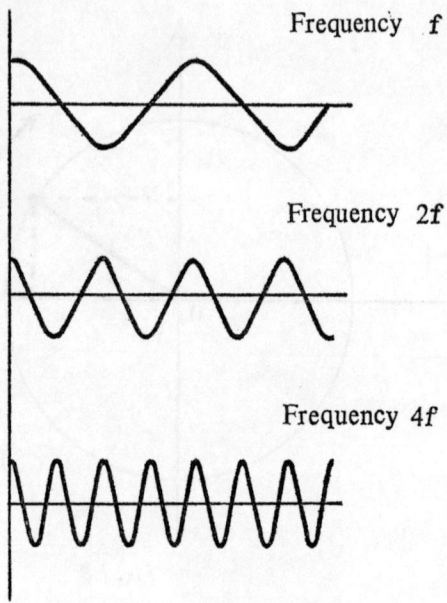

Fig. 1.4

useful to be able to start a cosine (or sine) wave part of the way through one period at $t = 0$. This is called the *phase* of the wave, and mathematically is ϕ in the expression $y = a \cos(\omega t + \phi)$.

Waves usually move or progress—or appear to progress; e.g. the sea wave moves towards the shore. This sort of effect is described mathematically by taking the basic wave shape; say $y = a \cos \omega x$ and letting the whole move with velocity v (see Fig. 1.5).

The wave shape still repeats itself every $x = 2\pi/\omega$ and the extreme values are still $\pm a$. The effect of the whole wave advancing with velocity v in the positive x direction is the same as moving the origin a distance vt after t seconds. The new x value is then $(x - vt)$ in terms of the original x value; and so the wave after time t is

$$y = a \cos [\omega(x - vt)].$$

A simple advancing wave like this has wavelength $2\pi/\omega$ as before.

4

CONCEPTS IN COMMUNICATIONS

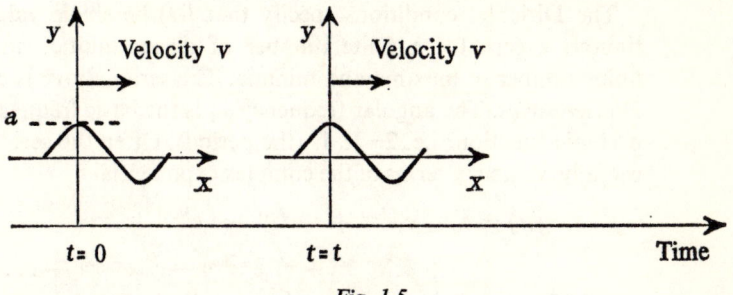

Fig. 1.5

The period here is the time it takes for the wave to move a distance of one wavelength,

$$T = \frac{2\pi}{\omega} \bigg/ v = \frac{2\pi}{\omega v}.$$

The frequency here is $\frac{1}{T} = f = \frac{\omega v}{2\pi}$

The relations are

$$\text{Velocity of wave} = \frac{\text{Wavelength}}{\text{Period}}$$

$$\text{Frequency} = \frac{1}{\text{Period}}$$

Waves can occur in many forms. There are sound waves in solids, liquids and gases; water waves; electric waves (waves in electric fields); magnetic waves (waves in magnetic fields); electromagnetic waves (combination of electric and magnetic effects). In each and every case frequency, period, phase, wavelength and amplitude are used to describe such waves.

1.3 Fourier analysis—frequency or harmonic analysis

The mathematician Fourier showed that any periodic function which satisfies certain conditions known as Dirichlet's conditions can be expanded as

$$f(t) = \frac{a_0}{2} + a_1 \cos \omega_1 t + a_2 \cos 2\omega_1 t + a_3 \cos 3\omega_1 t + \ldots$$
$$+ b_1 \sin \omega_1 t + b_2 \sin 2\omega_1 t + b_3 \sin 3\omega_1 t + \ldots$$

DATA TRANSMISSION

The Dirichlet conditions specify that $f(t)$ be single valued, continuous except for a finite number of discontinuities and have a finite number of maxima and minima. The series above is called the *Fourier series*. The angular frequency ω_1 is the basic frequency of the periodic function; i.e. $2\pi \times 1/$ (the period). Often the series is more usefully written in terms of the complex exponential

$$f(t) = c_0 + c_1 e^{i\omega_1 t} + c_2 e^{2i\omega_1 t} + c_3 e^{3i\omega_1 t} + \ldots$$
$$+ c_{-1} e^{-i\omega_1 t} + c_{-2} e^{-2i\omega_1 t} + c_{-3} e^{-3i\omega_1 t} + \ldots$$

because the coefficients $c_{\pm k}$ are easier to calculate, the series is more compact, and phases are obtained more easily.
The coefficients

$$a_k = \frac{2}{T} \int_0^T f(t) \cos(k\omega_1 t)\, dt \quad \text{for } k = 0, 1, 2 \ldots$$

$$b_k = \frac{2}{T} \int_0^T f(t) \sin(k\omega_1 t)\, dt \quad \text{for } k = 1, 2, 3 \ldots$$

or $\quad c_k = \frac{1}{T} \int_0^T f(t) e^{-ik\omega_1 t}\, dt \quad \text{for } k = 0, \pm 1, \pm 2, \ldots$

$T = $ wavelength or fundamental period, i.e. $T = 2\pi/\omega_1$. (See ref. 1.1.) Since $e^{ik\omega_1 t} = \cos(k\omega_1 t) + i \sin(k\omega_1 t)$ then either the trigonometric or the exponential Fourier series has a most important practical meaning. This is that any periodic function, for example a complicated wave, can be regarded as a superposition or combination of simple harmonic waves. For example, consider the periodic square wave shown in Fig. 1.6; this type of wave is very common in digital machines. Calculating the coefficients of the Fourier series gives

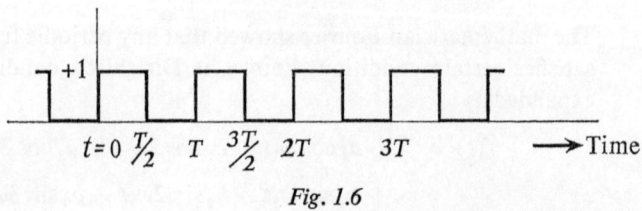

Fig. 1.6

$$a_0 = \frac{2}{T}\int_0^T f(t)\,dt = \frac{2}{T}\int_0^{\frac{1}{2}T} 1\cdot dt$$
$$= 1$$

since $f(t) = 1$ for $0 < t < \tfrac{1}{2}T$
$\qquad\quad\; = 0$ for $\tfrac{1}{2}T < t < T$

$$a_k = \frac{2}{T}\int_0^T f(t)\cos(k\omega_1 t)\,dt = \frac{2}{T}\int_0^{\frac{1}{2}T}\cos(k\omega_1 t)\,dt$$

$$= \frac{2}{T}\left[\frac{\sin(k\omega_1 t)}{k\omega_1}\right]_0^{\frac{1}{2}T} = \frac{2}{T}\left[\frac{\sin(k\omega_1 T/2)}{k\omega_1}\right]$$

but $\omega_1 = \dfrac{2\pi}{T}$ since the fundamental frequency $f_1 = \dfrac{1}{T}$

thus $a_k \quad = \dfrac{2}{T}\left[\dfrac{\sin\left(k\cdot\dfrac{2\pi}{T}\cdot\dfrac{T}{2}\right)}{k\dfrac{2\pi}{T}}\right]$

$$= \left[\frac{\sin(k\pi)}{k\pi}\right]$$

$= 0$ for all integer values of k;

that is, there are no cosine terms in the series.

Similarly
$$b_k = \frac{2}{T}\int_0^T f(t)\sin(k\omega_1 t)\,dt = \frac{2}{T}\int_0^{\frac{1}{2}T}\sin(k\omega_1 t)\,dt$$

$$= \frac{2}{T}\left[\frac{-\cos(k\omega_1 t)}{k\omega_1}\right]_0^{\frac{1}{2}T} = \frac{2}{T}\left[\frac{-\cos(k\omega_1\frac{1}{2}T)}{k\omega_1} - -\frac{1}{k\omega_1}\right]$$

as before using $\omega_1 = \dfrac{2\pi}{T}$

$$b_k = \frac{2}{T}\left[\frac{1-\cos(k\pi)}{k\dfrac{2\pi}{T}}\right] = \frac{1-\cos(k\pi)}{k\pi}$$

DATA TRANSMISSION

Thus $b_1 = 2/\pi$
$b_2 = 0$
$b_3 = 2/3\pi$
$b_4 = 0$
$b_5 = 2/5\pi$
etc.

So the Fourier Series for $f(t)$ is

$$\frac{1}{2} + \frac{2}{\pi}\left[\sin(\omega_1 t) + \frac{1}{3}\sin(3\omega_1 t) + \frac{1}{5}\sin(5\omega_1 t) + \ldots\right]$$

The complicated square wave can thus be written and used as a summation of simple sine functions whose frequencies are

$$\omega_1, 3\omega_1, 5\omega_1, \ldots, \text{etc}, \quad \text{where } \omega_1 = \frac{2\pi}{T}$$

The communications engineer uses two complementary representations for $f(t)$, one is the picture of the square wave as a function of

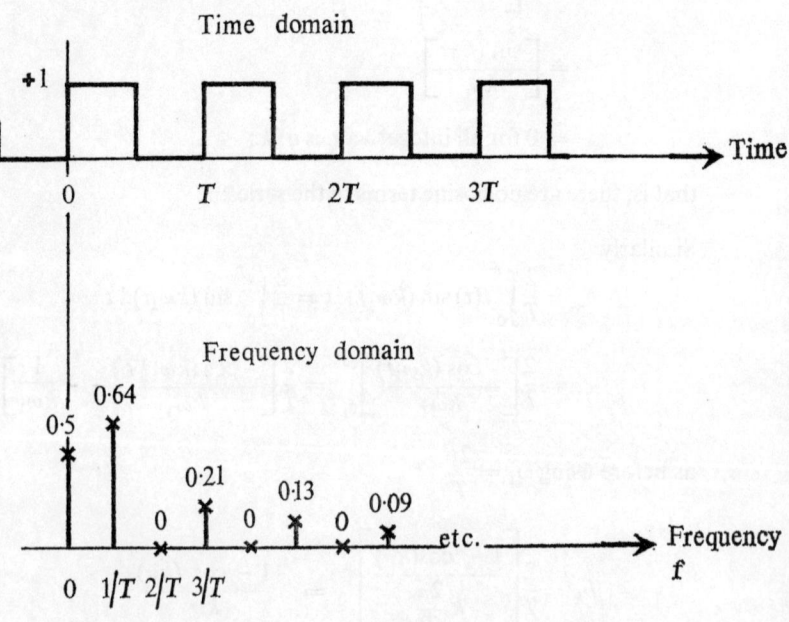

Fig. 1.7

time, the other is a sum of simple harmonic waves each with different amplitude, frequency and phase. One representation is said to be in the time domain, the other in the frequency domain as shown in Fig. 1.7. Since the frequency function consists of discrete frequencies, the picture in the frequency domain is a series of lines; it is called a line spectrum. The technique of analysing a time function into a frequency function is called harmonic analysis or sometimes frequency analysis. Many practical systems are more easily thought of in terms of frequency functions rather than time functions.

The Fourier series for periodic functions is usually written as the pair of identities

$$f(t) = \sum_{-\infty}^{+\infty} c_k e^{ik\omega_1 t}, \qquad c_k = \frac{1}{T} \int_0^T f(t) e^{-ik\omega_1 t} \, dt.$$

There is another pair that apply to non periodic (called aperiodic) functions, namely

$$f(t) = \int_{-\infty}^{+\infty} F(\omega) e^{i\omega t} d\omega, \qquad F(\omega) = \frac{1}{2\pi} \int_{-\infty}^{+\infty} f(t) e^{-i\omega t} \, dt.$$

These two equations are called the Fourier transform, and the combined integral

$$f(t) = \int_{-\infty}^{+\infty} \left[\frac{1}{2\pi} \int_{-\infty}^{+\infty} f(t') e^{-i\omega t'} \, dt' \right] e^{i\omega t} \, d\omega$$

is called the Fourier integral. The aperiodic function always gives rise to a continuum of frequencies in the frequency domain, called a continuous spectrum.

The conclusion is that any real, time function, periodic or aperiodic, can be analysed to a frequency function. For simplicity the operation of analysing is often symbolised by the letter F. Thus if $f(t)$ is any time domain function then $[F f(t)]$ or $F(\omega)$ is the equivalent frequency domain function.

The reader is referred to Chapter 1 of ref. 1.1 for further reading and extension of these topics.

The reason for the importance of frequency analysis is that the communications engineer usually describes his system or part of a system in terms of a frequency function. He says for example that the path between the transmitter and receiver has certain characteristics, and one of these characteristics is that only certain frequencies are

DATA TRANSMISSION

permitted to pass. The question that naturally arises is 'What happens to the square wave if the frequencies are restricted in some way?' Figure 1.8 shows the time domain picture for a few frequency combinations. This fact, that practical situations only permit a range of frequencies to be transmitted, is described by saying it is 'band limited'. The 'band' here is a band of frequencies. The band of allowed frequencies is sometimes called the bandwidth of the system. A frequency of 1 cycle per second is written 1 Hz, thus in Fig. 1.8, if $T = 1$ msec then $f_1 = 1000$ Hz, and the first four frequency components are contained in a band 0 to 4000 Hz.

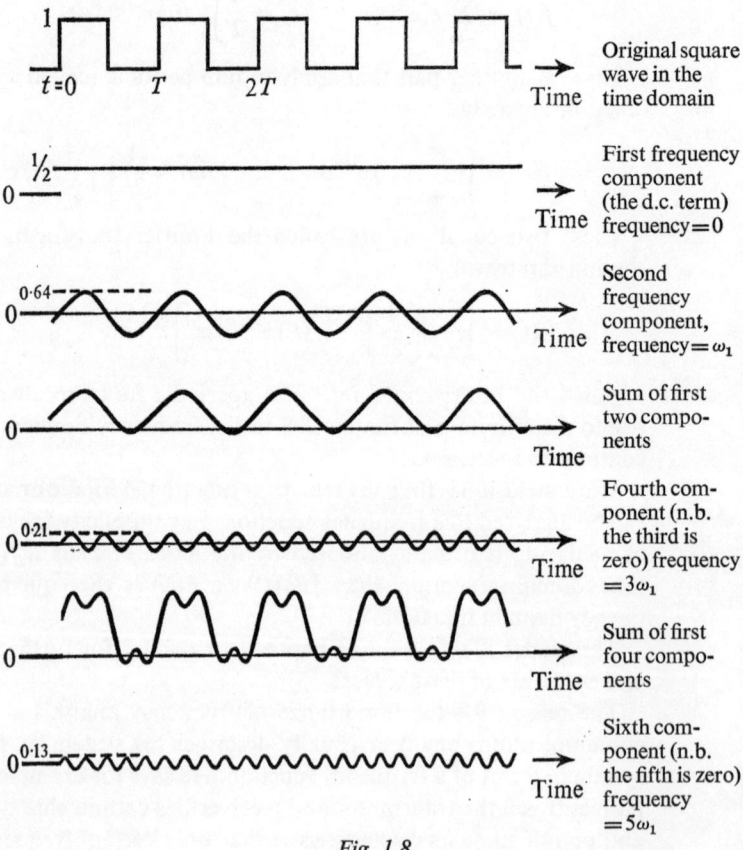

Fig. 1.8

For completeness we state briefly three further functions very useful in frequency analysis (see ref. 1.1).

(1) the Autocorrelation function defined for periodic $f_1(t)$ as

$$\phi_{11}(\tau) = \frac{1}{T} \int_0^T f_1(t).f_1(t+\tau)\,dt \quad \text{or for aperiodic } f_1(t) \text{ as}$$

$$\phi_{11}(\tau) = \int_{-\infty}^{+\infty} f_1(t).f_1(t+\tau)\,dt,$$

(2) the Crosscorrelation function defined for periodic $f_1(t)$ and $f_2(t)$ as

$$\phi_{12}(\tau) = \frac{1}{T} \int_0^T f_1(t).f_2(t+\tau)\,dt$$

or for aperiodic $f_1(t)$ and $f_2(t)$ as

$$\phi_{12}(\tau) = \int_{-\infty}^{+\infty} f_1(t).f_2(t+\tau)\,dt,$$

(3) the Convolution function defined for periodic $f_1(t)$ and $f_2(t)$ as

$$\rho_{12}(\tau) = \frac{1}{T} \int_0^T f_1(t).f_2(\tau - t)\,dt$$

or for aperiodic $f_1(t)$ and $f_2(t)$ as

$$\rho_{12}(\tau) = \int_{-\infty}^{+\infty} f_1(t).f_2(\tau - t)\,dt.$$

The importance of these functions is shown by taking their Fourier transforms. The convolution function in particular has great practical importance. If we denote F to mean the Fourier transform, and the symbol $*$ to mean convolution then

$$F[\rho_{12}(\tau)] = F[f_1(t) * f_2(t)]$$

can be shown to be proportional to

$$F[f_1(t)].F[f_2(t)].$$

In words it says that convolving in the time domain is equivalent to multiplication in the frequency domain. The converse is also true, that convolving in the frequency domain is equivalent to multiplication in the time domain. The usefulness of this is illustrated by considering the system $g_i(t) \to \boxed{h(t)} \to g_0(t)$. The function $g_i(t)$ is input to a

DATA TRANSMISSION

device whose transfer function is $h(t)$. The output is $g_0(t)$. By definition (of transfer functions)

$$F[g_0(t)] = F[g_i(t)].F[h(t)];$$

i.e. it multiplies in the frequency domain.
If $g_i(t)$ is a unit impulse then $F[g_i(t)] = 1$, which gives $F[g_o(t)] = F[h(t)]$ or $g_o(t) = h(t)$. Thus the transfer function, as a function of time, is given by the output function when the input is a unit impulse.

Now using the convolution theorem

$$F[g_0(t)] = F[g_i(t)].F[h(t)] = F[g_i(t)_*h(t)],$$

which gives $g_0(t) = g_i(t)_*h(t)$.
So that if $h(t)$ is known the output $g_0(t)$ can be calculated for any input $g_i(t)$ by convolving $g_i(t)$ with $h(t)$.

1.4 Information and channel capacity

1.4.1 Information

A communication system is one in which messages are passed between a transmitter and a receiver. To describe this system it is necessary to think of a message as information. Dictionary definitions of the word information involve meaning and this creates problems for the communications engineer. One of the beauties of the natural language is the arbitrariness or vagueness of words and sentences, and this conflicts with the scientists' requirements, which are preciseness and complete definability. To overcome these difficulties the communications engineer does not permit semantics to enter into the action of communicating. Semantics often affects how information is stored and retrieved but not the act of communicating, since what may be meaningful to one person may be nonsense to another.

The scientific definition of information is statistical. Consider a situation in which P different, but equally probable events, might occur. If some 'information' about the system is given then it can be specified that one of the events is more likely to occur, or even that only one of the events will occur. The larger P is the greater the

CONCEPTS IN COMMUNICATIONS

initial uncertainty and the more information is required to make a choice. The definition of the *information* is

$$I = K \log_e P,$$

where K is a constant. There are a number of good reasons for choosing the log function, the reader is referred to books on information theory; e.g. ref 1.2.

There seems to be some confusion about the units of information. Information as such is dimensionless, but it has become customary to express it in terms of bits (binary digits). In this case the situation of P equally probable events is fixed as W events, each being either 0 or 1. The total number of possibilities is thus $P = 2^W$, and the information $I = K \log_e 2^W = KW \log_e 2$. If it is required to express I as equivalent to a certain number of binary digits then that number must be W in this case. So to express I in bits, $KW \log_e 2$ must be equal to W, or $K = 1/\log_e 2$. Thus the definition of information could be

$$I = \frac{1}{\log_e 2} \cdot \log_e P \text{ bits} = \log_2 P \text{ bits}.$$

It is useful in some circumstances to relate the information to the decimal system. In this case the situation of P equally probable events is fixed as D events, each being either 0, 1, 2, 3, ... 9. The total number of possibilities is thus $P = 10^D$ and the information is $I = K \log_e 10^D = KD \log_e 10$. The name given to one decimal digit is a ban, and if it is required to express I as equivalent to a certain number of bans then that number must be D in this case. So to express I in bans, $KD \log_e 10 = D$, or, $K = 1/\log_e 10$.

Thus the definition of information could be

$$I = \frac{1}{\log_e 10} \cdot \log_e P \text{ bans,} = \log_{10} P \text{ bans}.$$

It is to be noted that I expressed in bits or bans (or any other system) merely fixes a value for the constant K. As an example consider the amount of information in an English language sentence. It has an alphabet of 27 symbols (26 letters plus a space). A sentence of N equally probable symbols therefore contains

$$I = \log_2(27^N) \text{ bits} = N \log_2 27 \text{ bits}$$

DATA TRANSMISSION

or equivalently

$$\log_{10}(27^N) \text{ bans} = N \log_{10} 27 \text{ bans}.$$

The information per letter is, in this case,

$$I = \log_2 27 \text{ bits} = 4 \cdot 76 \text{ bits}$$

or $\log_{10} 27$ bans $= 1 \cdot 43$ bans.

However, this figure is not correct, since in practice each letter does not occur with equal probability. Shannon, ref. 1.3, showed that the average information per symbol is

$$I = -K \sum_{j=1}^{M} p_j \log_e p_j,$$

where p_j is the probability of the jth symbol, and M is the total number of symbols.

Note that the system discussed above has a situation of P equally probable events, then

$$p_1 = p_2 = p_3 \ldots = p_M = \frac{1}{P}$$

and Shannon's formula

is
$$I = -K \sum_{1}^{M} p_j \log_e p_j$$

$$I = -K \sum_{1}^{P} \frac{1}{P} \log_e \frac{1}{P} = K \log_e P,$$

which agrees with the original definition.

If the set of single letter probabilities for English language is used then Shannon's formula gives the information as 4·03 bits per letter. But this is not correct, since letters do not occur independently; they are correlated. The fact that 'TH' has already been received means that the further reception of 'E' does not give 4·03 bits of information, but considerably less since it is highly likely. Shannon calls this correlation redundancy. Further work seems to indicate that, in fact, the amount of information is about 1·4 bits per letter. The reader is referred to books on information theory, and also section 5.2.

Note that the information as a number of bits per symbol does not say whether a collection of symbols has any meaning. For the communications engineer information is a practical measure of the number of possible events he must cater for in his transmission system.

1.4.2 Channel capacity

A channel is the medium for the transmission of a message; the *capacity* of that channel is described in terms of the information it can transmit and not the number of symbols. The usual definition of channel capacity is the maximum rate at which information can be transmitted without error, in units of bits per second.

Provided the information in a message is matched to its channel then in the absence of noise that message will arrive without error.

Very early in the history of telegraphy it was known that the rate at which data can be transmitted is limited. The rate can be shown to be proportional to the bandwidth of frequencies that are allowed on the channel. (See ref. 1.4.)

Let symbols be transmitted at intervals of time b. Then a minimum band of frequencies $\frac{1}{2}b$ is required to transmit them (see section 4.2 and references 1.5 and 1.6). Alternatively, if a channel is capable of transmitting a bandwidth of frequencies W (which equals $1/b$) then it can transmit symbols at a rate up to $2W$, but not greater. This is really a restatement of the sampling theorem. (See section 4.2.)

So using the definition of channel capacity, that it is the maximum information rate, then

$$C = \frac{I}{T} = \frac{K}{T}\log_e P = \frac{1}{T} \cdot \log_2 P \text{ bits per sec,}$$

where $1/T$ is the rate of transmission of the symbols. If the number of possible events P are equated with L the number of distinct channel signal levels, then since $(1/T)_{max} = 2W$, $C = 2W \log_2 L$ bits per second. This is the capacity of a noiseless channel. Alternatively put, it says that to transmit information at C bits per second along a noiseless channel requires a bandwidth of

$$W = \frac{C}{2 \log_2 L} \text{ Hz.}$$

If there is noise present, which is the case in practice (see section 2.3), then the very important relation derived by Shannon (see ref. 1.7) is that

$$C = W \log_2\left(1 + \frac{S}{N}\right) \text{bits per sec,}$$

where W is the bandwidth, S the signal power and N the noise power for the channel. The noise in this case must be white (or Gaussian)—see section 5.1.

Any channel therefore has a *maximum* rate at which information can be transmitted. This *maximum* rate is based upon statistical analysis very similar to that in thermodynamics. The equation for the channel capacity is very similar to that for entropy (in fact negative entropy—see ref. 1.2), which is linked to one of the fundamental laws of physics—the Second Law of Thermodynamics. The building of a transmission system which operates at any information rate greater than that given by Shannon's formula is equivalent to building a perpetual motion machine—that is, impossible.

Communications engineers usually refer to signal levels in terms of decibels, a relative measure. The number of decibels (or dB's) is equal to $10 \log_{10} (B_1/B_2)$, where B_1 and B_2 are the powers in the two signals being compared. So taking as an example a telephone line, a common signal to noise ratio is 20 dB, which means that

$$20 = 10 \log_{10}\left(\frac{S}{N}\right),$$

where S is the signal power and N the noise power. So $\frac{S}{N} = \frac{100}{1}$. If the bandwidth of a telephone line is assumed to be 3000 Hz, then Shannon's formula gives

$$C = 3000 \log_2\left(1 + \frac{100}{1}\right) \text{ bits per second}$$

$$= 3000 \times \frac{\log_{10}(101)}{\log_{10}(2)}$$

$$= \frac{3000 \times 2{\cdot}00432}{0{\cdot}30103}$$

$$= 19{,}963 \text{ bits per second}.$$

This is the maximum information rate for signals using the line. Making the bandwidth wider, or the signal-to-noise ratio larger, will increase the channel capacity.

In practice, transmission on telephone lines works at a very much lower information rate, for example 600 or 1200 bits per second, and 2400 bits per second on a good quality telephone line. The reasons

CONCEPTS IN COMMUNICATIONS

for this are complicated; the actual bandwidth may be much less than 3000 Hz; the noise is not Gaussian (see section 5.1), nor is it constant at 20 dB signal/noise ratio; to achieve high information rates demands complicated coding which may not be used for many reasons.

1.5 Transmission media

In the early days of telegraphy the communication channel used was a pair of copper wires separated by an air gap by being suspended on poles. They are still to be seen in some countries. The development of these *open pairs* into cable form was made by using insulating materials such as paper. Nowadays a telephone cable consists of hundreds of these wire pairs packed together, insulated with materials such as P.V.C., with an outer protective covering of lead (an outer diameter of 5 centimetres is typical). The other early form of communicating was the wireless radio wave broadcast, the channel being an air path.

There are practical problems associated with the transmission of high-frequency signals along wire conductors. A specially constructed wire developed for high (radio) frequencies is the coaxial cable, which consists of a solid conducting core surrounded by an insulator which itself is completely surrounded by a conducting screen. Typically a telephone cable will consist of about six such coaxial cables packed together with several hundred normal telephone wire pairs.

High-frequency waves can be transmitted over an 'air' path. At the very high radio frequencies (e.g. 10^{10} Hz) a microwave system is used with special transmitting and receiving antennae. These are built as a series of 'line-of-sight' paths, typically inter-city.

The use of a satellite as a relay station for the wireless radio wave transmission has enabled large distances to be covered with fewer earth stations.

To improve the information rate demands higher frequencies. Frequencies above about 10^{11} Hz can use devices called waveguides to carry them. The optical waveguide is one using visible light frequencies of about 10^{14} Hz. For further reading see reference 5.5.

2 Communication codes

In this section various existing codes will be discussed. Telegraphic transmission is in serial asynchronous form (see section 6.3). The telegraph machines that were developed to transmit and/or receive and print characters automatically almost universally work in a two-state or binary mode. The information or message is transmitted as a sequence of characters, each character being in a particular code.

2.1 The Baudot code

The most common, and one of the earliest, codes is the Baudot 5-bit telegraph code (see Fig. 2.1). Five bits are allocated per character. Since 5 bits would permit only 32 different characters, the range is extended by having an upper- and a lower-case form; these two forms are indicated by two special characters, the letter shift and the figure shift. When a letter shift is sent all the characters that follow it are in letter case until a figure shift is sent. There are four other special non-printing characters, blank, space, carriage-return and line-feed. These six special characters are recognised in either upper or lower case, and are sometimes called *stunt* characters.

The CCITT International Alphabet number 2 has been adopted by most countries in the world, but there are some minor differences particularly in the U.S.A.—see Fig. 2.1.

2.2 The ASCII code

The five bit code above is widely used for telegraph (Telex) systems; but for use with computer-type machines it imposes too many restrictions; e.g. limited character set. In data transmission, particularly with computers, it is obviously sensible to standardise on the

COMMUNICATION CODES

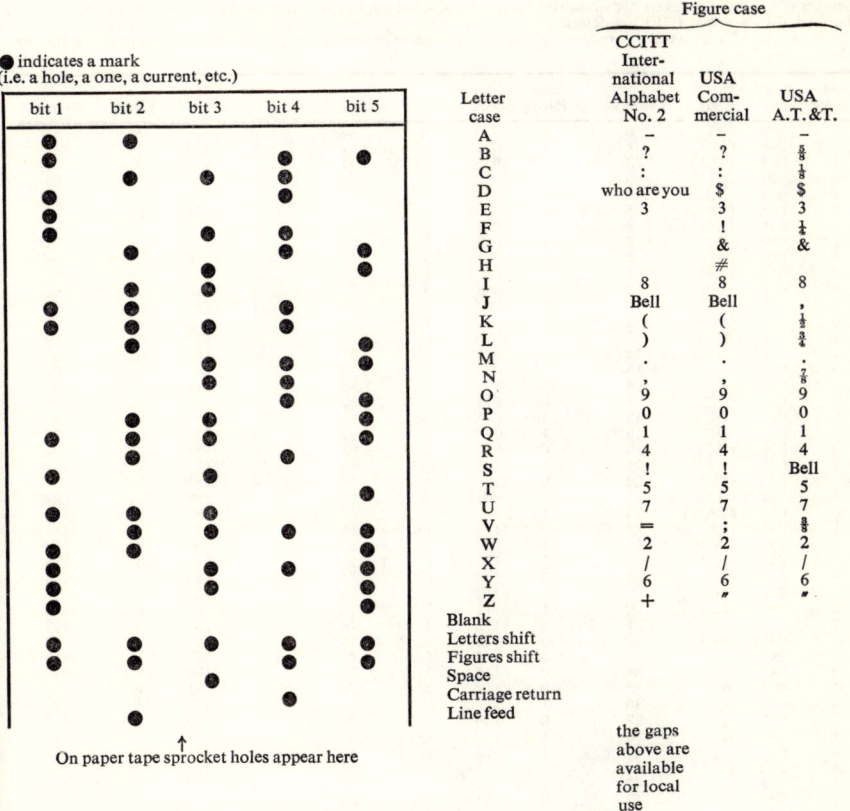

Fig. 2.1 Baudot 5-unit telegraph code

code that is used. There is widespread acceptance of the ASCII code (American Standard Code for Information Interchange; see ref. 2.1), although some manufacturers prefer their own (see later). The ASCII code is defined with seven bits; some users transmit it as an eight bit code with the eighth bit always a one; many users incorporate the eighth bit as a parity digit—see section 5.2. This code is very similar to the International Standards Organisation Code 7 (ISO 7).

Figure 2.2 gives the ASCII code.

DATA TRANSMISSION

Groups of letters in brackets are non-printing control characters, in particular: (CR) = carriage return, (LF) = line feed, (SP) = space, (DEL) = delete.
Codes 140_8 to 176_8 are available only on dual-case machines. On single-case machines they print as codes 100_8 to 136_8.

Even Parity	7 bit octal	Char.	Decimal	Even Parity	7 bit octal	Char.	Decimal	Even Parity	7 bit octal	Char.	Decimal
0	000	(NUL)	0	0	053	+	43	0	126	V	86
1	001	(SOH)	1	1	054	,	44	1	127	W	87
1	002	(STX)	2	0	055	−	45	1	130	X	88
0	003	(ETX)	3	0	056	.	46	0	131	Y	89
1	004	(EOT)	4	1	057	/	47	0	132	Z	90
0	005	(ENQ)	5	0	060	ø	48	1	133	[	91
0	006	(ACK)	6	1	061	1	49	0	134	/	92
1	007	(BEL)	7	1	062	2	50	1	135	]	93
1	010	(BS)	8	0	063	3	51	1	136	↑	94
0	011	(HT)	9	1	064	4	52	0	137	←	95
0	012	(LF)	10	0	065	5	53	0	140	∧	96
1	013	(VT)	11	0	066	6	54	1	141	a	97
0	014	(FF)	12	1	067	7	55	1	142	b	98
1	015	(CR)	13	1	070	8	56	0	143	c	99
1	016	(SO)	14	0	071	9	57	1	144	d	100
0	017	(SI)	15	0	072	:	58	0	145	e	101
1	020	(DLE)	16	1	073	;	59	0	146	f	102
0	021	(DC1)	17	0	074	<	60	1	147	g	103
0	022	(DC2)	18	1	075	=	61	1	150	h	104
1	023	(DC3)	19	1	076	>	62	0	151	i	105
0	024	(DC4)	20	0	077	?	63	0	152	j	106
1	025	(NAK)	21	1	100	@	64	1	153	k	107
1	026	(SYN)	22	0	101	A	65	0	154	l	108
0	027	(ETB)	23	0	102	B	66	1	155	m	109
0	030	(CAN)	24	1	103	C	67	1	156	n	110
1	031	(EM)	25	0	104	D	68	0	157	o	111
1	032	(SUB)	26	1	105	E	69	1	160	p	112
0	033	(ESC)	27	1	106	F	70	0	161	q	113
1	034	(FS)	28	0	107	G	71	0	162	r	114
0	035	(GS)	29	0	110	H	72	1	163	s	115
0	036	(RS)	30	1	111	I	73	0	164	t	116
1	037	(US)	31	1	112	J	74	1	165	u	117
1	040	(SP)	32	0	113	K	75	1	166	v	118
0	041	!	33	1	114	L	76	0	167	w	119
0	042	"	34	0	115	M	77	0	170	x	120
1	043	#	35	0	116	N	78	1	171	y	121
0	044	$	36	1	117	O	79	1	172	z	122
1	045	%	37	0	120	P	80	0	173	{	123
1	046	&	38	1	121	Q	81	1	174	|	124
0	047	'	39	1	122	R	82	0	175	}	125
0	050	(	40	0	123	S	83	0	176	~	126
1	051	)	41	1	124	T	84	1	177	(DEL)	127
1	052	*	42	0	125	U	85				

Fig. 2.2. ASCII code

2.3 Binary and binary coded decimal (BCD)

As data transmission frequently involves only numbers, it is often convenient to transmit these numbers in their binary form rather than convert them to the ASCII code.

The problem of converting large decimal numbers to binary is overcome if instead of the whole decimal number being converted each digit in turn is converted and transmitted sequentially.

To represent the decimal digits 0, 1, . . . 9 four bits are required for each digit; if each decimal digit is converted to binary then this is one

COMMUNICATION CODES

form of Binary Coded Decimal. For example the decimal number

1 2 3 4 5 6 7

in binary is

100101101011010000111

and in binary coded decimal is

0001 0010 0011 0100 0101 0110 0111

There are a number of forms of Binary Coded Decimal; Fig. 2.3 gives some of them. They have useful properties for easy operation in analog/digital converters. For example, the Excess Three Code has the 1's complement of a character equal to its 9's complement in decimal; the Gray Code has adjacent characters differing by only one binary digit.

Decimal digit	The 8421 or Straight Binary	The 2421 Code	The Excess Three Code	The Gray Code
0	0000	0000	0011	0000
1	0001	0001	0100	0001
2	0010	0010	0101	0011
3	0011	0011	0110	0010
4	0100	0100	0111	0110
5	0101	0101	1000	0111
6	0110	0110	1001	0101
7	0111	0111	1010	0100
8	1000	1110	1011	1100
9	1001	1111	1100	1101

Fig. 2.3. Binary coded decimal

2.4 The EBCDIC code

EBCDIC stands for Extended Binary Coded Decimal Interchange Code. Each character is represented by one eight-bit number; the eight bits are divided into two groups of 4. Since 4 bits gives decimal 0 through 15 the single letters A, B, C, D, E and F are used for the numbers 10, 11, 12, 13, 14 and 15. This gives a true hexadecimal code. An EBCDIC character is therefore a pair of hexadecimal digits; for

DATA TRANSMISSION

example, hexadecimal C1 is the code for the letter A; C1 is 1100 0001 in binary coded hexadecimal, or 193 in decimal, or 11000001 in binary.

Figure 2.4 gives the set of EBCDIC characters, together with other possible meanings.

2.5 Control characters

In both the ASCII and EBCDIC codes some of the characters have special meanings; they are used as special control characters. For example, serial synchronous transmission (see Chapter 10) requires characters for message control such as NAK, ENQ, ETX, SYN. The ASCII code allocates one character for each of these control characters (see Fig. 2.2). The ASCII character 006 is the code for ACK. The EBCDIC character 37 is the code for EOT (End of Transmission). The provision of single control characters in the codes greatly simplifies the hardware in, for example, the serial synchronous line interface. The interface then has only to look for single special characters. The meaning of some of the ASCII control characters is as follows:

SOH   Start of Header. Used at the beginning of a sequence of characters to indicate an address. Terminated by STX.
STX   Start of Text. Used at the beginning of a sequence of characters which are the text.
ETX   End of Text.
EOT   End of Transmission.
ENQ   Enquiry. Used to request a response, which typically is the address, status, etc, of the receiver.
ACK   Acknowledge. Indicates successful reception.
NAK   Negative Acknowledge. Indicates unsuccessful reception.
SYN   Synchronisation. The synchronisation character.
ETB   End of Transmission block. Terminates characters headed by SOH or STX.

The way in which these control characters are used is described in Chapter 10.

COMMUNICATION CODES

2 bit hexadec.	Character	Decimal	2 bit hexadec.	Character	Decimal	2 bit hexadec.	Character	Decimal
00	(null)	0	38		56	71		113
01		1	39		57	72		114
02		2	3A		58	73		115
03		3	3B		59	74		116
04		4	3C		60	75		117
05	(tab)	5	3D		61	76		118
06		6	3E		62	77		119
07	(delete)	7	3F		63	78		120
08		8	40	(space)	64	79		121
09		9	41		65	7A	:	122
0A		10	42		66	7B	@	123
0B		11	43		67	7C	'	124
0C		12	44		68	7D	=	125
0D		13	45		69	7E	"	126
0E		14	46		70	7F		127
0F		15	47		71	80		128
10		16	48		72	81	a	129
11		17	49		73	82	b	130
12		18	4A	£	74	83	c	131
13		19	4B	.	75	84	d	132
14		20	4C	<	76	85	e	133
15	(new line)	21	4D	(	77	86	f	134
16		22	4E	+	78	87	g	135
17		23	4F	ǀ	79	88	h	136
18		24	50	&	80	89	i	137
19		25	51		81	8A		138
1A		26	52		82	8B		139
1B		27	53		83	8C		140
1C		28	54		84	8D		141
1D		29	55		85	8E		142
1E		30	56		86	8F		143
1F		31	57		87	90		144
20		32	58		88	91	j	145
21		33	59		89	92	k	146
22	(field separator)	34	5A	!	90	93	l	147
23		35	5B	$	91	94	m	148
24		36	5C	*	92	95	n	149
25	(line feed)	37	5D	)	93	96	o	150
26		38	5E	;	94	97	p	151
27		39	5F	¬	95	98	q	152
28		40	60	—	96	99	r	153
29		41	61	/	97	9A		154
2A		42	62		98	9B		155
2B	(tab)	43	63		99	9C		156
2C		44	64		100	9D		157
2D	(carr. ret.)	45	65		101	9E		158
2E		46	66		102	9F		159
2F		47	67		103	A0		160
30		48	68		104	A1		161
31		49	69		105	A2	s	162
32		50	6A	^	106	A3	t	163
33		51	6B	,	107	A4	u	164
34		52	6C	%	108	A5	v	165
35		53	6D	—●—	109	A6	w	166
36		54	6E	>	110	A7	x	167
36		54	6F	?	111	A8	y	168
37	(EOT)	55	70		112	A9	z	169

Fig. 2.4 EBCDIC code

DATA TRANSMISSION

2 bit hexadec.	Character	Decimal	2 bit hexadec.	Character	Decimal	2 bit hexadec.	Character	Decimal
AA		170	C7	G	199	E4	U	228
AB		171	C8	H	200	E5	V	229
AC		172	C9	I	201	E6	W	230
AD		173	CA		202	E7	X	231
AE		174	CB		203	E8	Y	232
AF		175	CC		204	E9	Z	233
B0		176	CD		205	EA		234
B1		177	CE		206	EB		235
B2		178	CF		207	EC		236
B3		179	D0		208	ED		237
B4		180	D1	J	209	EE		238
B5		181	D2	K	210	EF		239
B6		182	D3	L	211	F0	0	240
B7		183	D4	M	212	F1	1	241
B8		184	D5	N	213	F2	2	242
B9		185	D6	O	214	F3	3	243
BA		186	D7	P	215	F4	4	244
BB		187	D8	Q	216	F5	5	245
BC		188	D9	R	217	F6	6	246
BD		189	DA		218	F7	7	247
BE		190	DB		219	F8	8	248
BF		191	DC		220	F9	9	249
C0		192	DD		221	FA		250
C1	A	193	DE		222	FB		251
C2	B	194	DF		223	FC		252
C3	C	195	E0	(blank)	224	FD		253
C4	D	196	E1		225	FE		254
C5	E	197	E2	S	226	FF		255
C6	F	198	E3	T	227			

Fig. 2.4 (contin.)

2.6 Other codes

There are many other codes in use. One very common example is the Friden Flexowriter Code. This is a six-bit plus parity and control code, used on electric typewriters.

Until fairly recently most computer manufacturers had their own individual character code set; these are still to be found in use.

3 Modulation

3.1 Introduction

It has been shown that any signal is made up of a range of frequencies of varying amplitude and phase. Speech, for example, is adequately covered by the frequency range 100 Hz to 4000 Hz. Other signals, depending upon their complexity, will require either more or fewer frequencies. The discovery in about 1890 by Hertz that electrically produced waves could be transmitted and received without being carried by wires was of great importance. It is called wireless transmission. It led to the practice of one transmitter broadcasting to a number of receivers without the need for wire conductor channels. In principle speech converted to an electrical signal can be amplified, fed to some aerial system and broadcast—that is, transmitted. Although there are practical difficulties about building such a system, there is, in fact, an even more important logical objection. The broadcasting of baseband speech (i.e. the actual speech frequencies) —or any other signal—means that within one reception area only one transmitter can be operating at any one time. If two or more

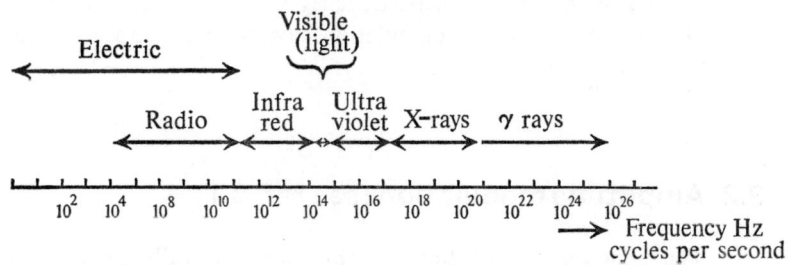

The electromagnetic spectrum

Fig. 3.1

DATA TRANSMISSION

baseband signals arrive at a receiver how is it to differentiate between them? Modulation was introduced as a solution to this and other problems. Basically it transforms the range of frequencies in the signal into another set at a different position in the frequency band —usually much higher. The signal is transmitted typically at frequencies the order of 10^8 Hz called radio frequencies (see Fig. 3.1).

The logical objection above is overcome by transmitting different signals in different parts of the frequency domain; the receiver is able to select any one particular signal by adjusting electrical circuits. Another advantage is that the dimensions of an aerial array need to be the order of the wavelength of the signal being transmitted (or received) if the aerial is to operate efficiently. Since baseband speech aerials would need dimensions of the order of $3 \times 10^8/1000$ metres = 300,000 metres, and frequencies of 10^8 would need dimensions of the order of $3 \times 10^8/10^8 = 3$ metres the advantage of operating at radio frequencies is apparent.

Modulation methods therefore were used exclusively at first in wireless, radio wave broadcasts to improve the efficiency of transmission, and to enable a receiver to choose between a number of transmitters broadcasting simultaneously.

A line or wire path has certain fixed electrical characteristics. It often happens that the range of frequencies it can accommodate is far greater than the range of frequencies in a signal to be transmitted along it. Modulation methods are also used in such cases to enable two or more signals to share the line.

In the early days of wireless broadcasting three types of modulation were devised. These will be discussed in this section. Since the Second World War, and particularly with the advent of solid-state circuits, other forms of modulation have been developed. These will be discussed in Chapter 4.

3.2 Amplitude modulation (see ref. 3.1)

The unmodulated continuous wave signal is fully represented by $y = A \cos(\omega t + \phi)$, where A is amplitude and ω is the angular frequency, i.e. frequency $f = \omega/2\pi$. Note, no great distinction will be made between ω and f, ω will always be angular frequency in radians per second, and f will be frequency in cycles per second, or Hz. ϕ is the phase or signal position relative to $t = 0$.

Consider, for simplicity, a wave

$$y_1 = A_1 \cos(\omega_1 t).$$

Amplitude modulation is the changing of the amplitude A_1 by some known function which will be the message. Consider changing A_1 by adding to it another complete wave $y_2 = A_2 \cos(\omega_2 t)$.

Then
$$\begin{aligned}y &= (A_1 + y_2) \cos(\omega_1 t) \\ &= (A_1 + A_2 \cos(\omega_2 t)) \cos(\omega_1 t) \\ &= A_1 \cos(\omega_1 t) + A_2 \cos(\omega_2 t) \cos(\omega_1 t) \\ &= A_1 \cos(\omega_1 t) + \tfrac{1}{2} A_2 \cos((\omega_1 + \omega_2) t) \\ &\quad + \tfrac{1}{2} A_2 \cos((\omega_1 - \omega_2) t)\end{aligned}$$

This contains three frequencies, one at the original angular frequency ω_1, one at $\omega_1 + \omega_2$ and one at $\omega_1 - \omega_2$. (see Fig. 3.2 (a).) y_1 is said to be amplitude modulated by y_2. The ratio A_2/A_1 is called the *modulation index* (or *factor*). In practice A_2 is less than A_1, since if $A_2 > A_1$ (giving a modulation index greater than 1) the resulting wave y will have more peaks than the modulating signal y_2. This situation leads to great practical problems in recovering the original signal y_2. (See Fig. 3.2(b).)

If the modulating signal y_2 consists of several frequencies then each one in y_2 gives rise to two in y as well as the single frequency ω_1. The frequency domain picture in this case is described as consisting of the frequency ω_1, with sidebands, a lower sideband containing frequencies of the form $\omega_1 - \omega_i$ and an upper sideband containing frequencies of the form $\omega_1 + \omega_i$, where $i = 2, 3, \ldots N$ for a signal y_2 with $(N-1)$ frequencies in it. The frequency ω_1 of y_1 always remains whilst those in y_2 will be shifted. Since the signal y_2 modulates the signal y_1, the signal y_2 is, in a sense, 'carried' by y_1. Because of this y_1 is called the *carrier wave*, and ω_1 the *carrier frequency*.

If ω_1 is a typical radio frequency, say $2\pi 10^8$ rads per sec and ω_2 a typical baseband frequency; say $2\pi \times 10^3$ rads per sec, then it can be seen that the spread of frequencies $\omega_1 - \omega_2$ to $\omega_1 + \omega_2$ is very limited. This fact, that the actual *bandwidth* is very narrow, means that in the radio frequency range many amplitude modulated signals can be accommodated. The different position of each signal y is

DATA TRANSMISSION

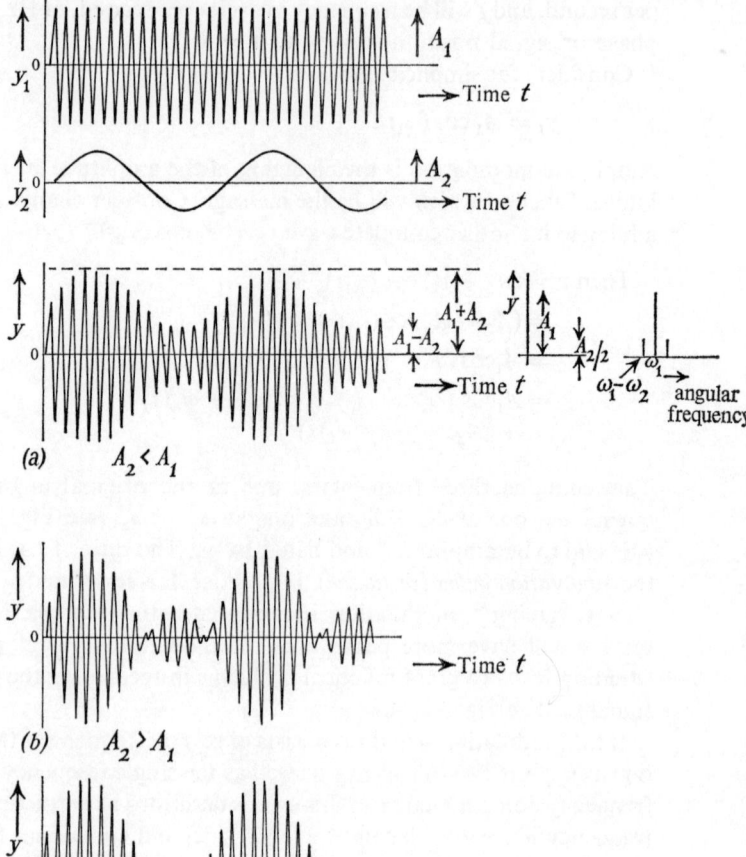

Fig. 3.2

28

MODULATION

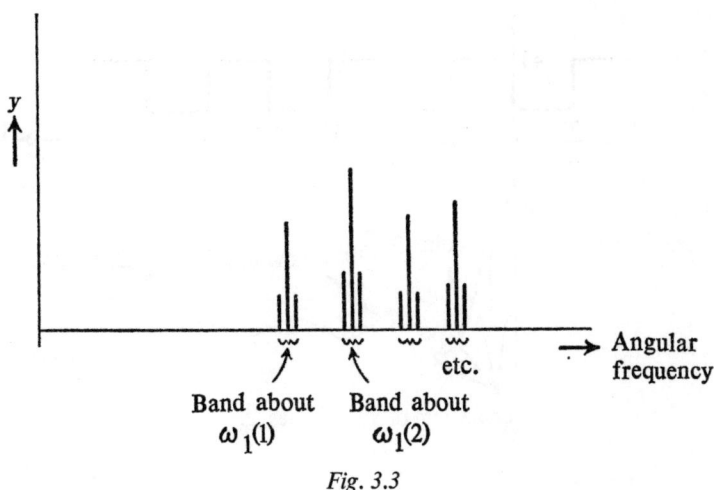

Fig. 3.3

determined by the actual value of ω_1. (See Fig. 3.3.) The value of ω locates, in the frequency domain, the position of a particular signal y. Therefore amplitude modulation is a method of conveying information, the message being the signal y_2 carried by the wave y_1. Many different messages can be transmitted at the same time by using a different carrier frequency for each message, an example of this is the domestic radio a.m. band.

Suppose that the message to be transmitted is a sequence of binary digits; for simplicity consider it to be of the form . . . 1 0 1 0 1 0 1 0, etc. Fourier analysis of this using the complex exponential form shows that the baseband signal is made up of a set of simple frequencies whose amplitudes vary like the $\sin(x)/x$ function—Fig. 3.4 illustrates two such signals; see also section 1.3.

The problem with this function is that strictly there is an infinity of frequencies in the baseband signal. When using amplitude modulation to carry it the bandwidth of the resulting wave is therefore theoretically infinite. In practice, though, since the amplitude of each frequency is getting progressively smaller, about 10 are regarded as sufficient to represent the message fairly accurately, thus the bandwidth for a 1000 Hz Square wave (i.e. $T = 1$ msec in Fig. 3.4(*a*)) restricted to two times $(10 \times 1/T) = 20,000$ Hz will enable such a wave to be recovered as a message.

29

DATA TRANSMISSION

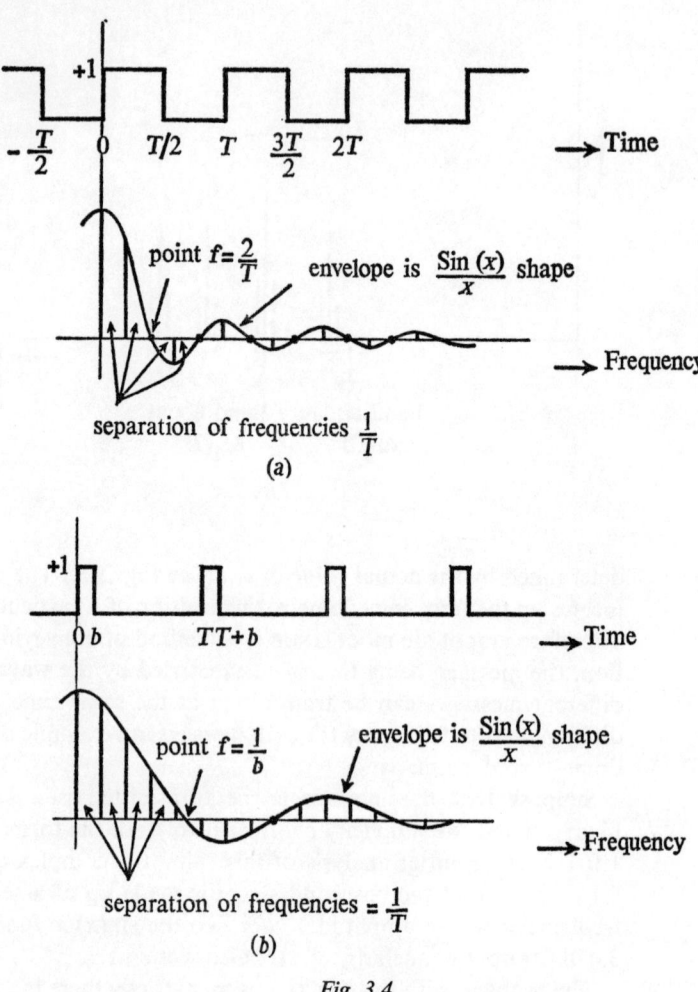

Fig. 3.4

It can be seen that two carrier frequencies should not be permitted to be so close to one another that the bands of frequencies resulting from amplitude modulation overlap. If this happens it causes interference, which for example is heard on the domestic radio a.m. band as one broadcasting station faintly in the background whilst one is listening to another—see also Chapter 5.

In order to pack the frequency domain efficiently often only one sideband is transmitted. This is called *single sideband amplitude modulation*. Also in order to utilise the transmitting power more efficiently it is common to find the carrier frequency removed before transmission. This is called *suppressed carrier amplitude modulation*. In the latter case the receiver must know the precise value of the carrier frequency, and use this information to recover the message.

3.3 Frequency modulation

Amplitude modulation is widely used as a method of superimposing a message signal on to a carrier signal. The main problem is that it is badly affected by noise, since any extraneous signal can simply add to it.

Frequency modulation was invented around the beginning of this century. Strangely, there is heated debate over who first suggested it. See ref. 3.2.

Frequency modulation gives greatly improved performance in the presence of noise. As the name implies, it is a method of superimposing the message signal y_2 on to the frequency of the carrier signal y_1, instead of on to the amplitude. There are practical difficulties about implementing such a system reliably, difficulties only overcome since approximately the 1940's.

As before, for simplicity, let the carrier wave be

$$y_1 = A_1 \cos(\omega_1 t)$$

Consider changing the frequency by adding to it another complete wave

$$y_2 = A_2 \cos(\omega_2 t)$$

Then

$$y = A_1 \cos((\omega_1 + y_2)t)$$
$$y = A_1 \cos((\omega_1 + A_2 \cos(\omega_2 t))t).$$

Since A_2 is the maximum absolute value of y_2, it is therefore the maximum absolute value in the deviation of the frequency. This is a complicated function. However in the solution of a commonly occurring differential equation

$$x^2 \frac{d^2 u}{dx^2} + x \frac{du}{dx} + (x^2 - n^2)u = 0$$

DATA TRANSMISSION

(where n is a constant) called Bessel's differential equation there arise the functions

$$J_n(x) = \sum_{k=0}^{\infty} \frac{(-1)^k}{k!\,(n+k)!} \left(\frac{x}{2}\right)^{n+2k}$$

which are called Bessel Functions of the First Kind. They obey the recurrence relation

$$2n\, J_n(x) = x J_{n-1}(x) + x J_{n+1}(x).$$

Examination of them shows that (see ref. 3.3)

$$\cos(x \cos(\phi)) = J_0(x) - 2 J_2(x) \cos(2\phi) + 2 J_4(x) \cos(4\phi) \ldots$$

and

$$\sin(x \sin(\phi)) = 2 J_1(x) \sin(\phi) - 2 J_3(x) \sin(3\phi) + 2 J_5(x) \sin(5\phi) \ldots$$

It can be shown (see ref. 3.4) that the frequency modulated function y can therefore be written as

$$y = A_1 J_0 \left(\frac{A_2}{\omega_2}\right) \cos(\omega_1 t)$$

$$+ A_1 J_1 \left(\frac{A_2}{\omega_2}\right) \Big[\cos((\omega_1 + \omega_2)t) - \cos((\omega_1 - \omega_2)t) \Big]$$

$$- A_1 J_2 \left(\frac{A_2}{\omega_2}\right) \Big[\cos((\omega_1 + 2\omega_2)t) - \cos((\omega_1 - 2\omega_2)t) \Big]$$

$$+ A_1 J_3 \left(\frac{A_2}{\omega_2}\right) \Big[\cos((\omega_1 + 3\omega_2)t) - \cos((\omega_1 - 3\omega_2)t) \Big] \ldots \text{etc.}$$

In other words, if the modulating function is a single frequency wave it gives rise to an infinite number of frequencies at $\omega_1 \pm \omega_2$, $\omega_1 \pm 2\omega_2$, $\omega_1 \pm 3\omega_2$ etc. The amplitude of each is given by the Bessel Functions $J_n(x)$; these approach zero rapidly when $n > (A_2/\omega_2)$. Thus if A_2/ω_2 is kept small the transmission can take place over a relatively narrow band of frequencies about ω_1; on the other hand since the message information is contained in these frequencies A_2/ω_2 must be large enough to give a sufficient number.

The factor A_2/ω_2 is called the *modulation index*. Practical values are usually the order of 10. Fig. 3.5 shows four possible spectra, and a typical f.m. wave. The frequency modulated signal is transmitted at

MODULATION

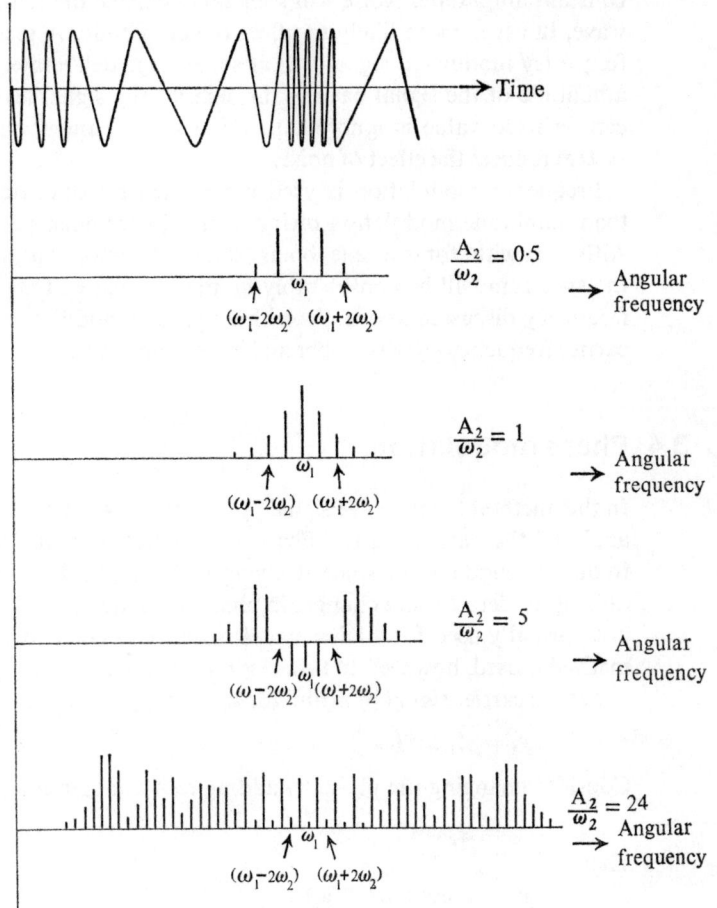

Frequency modulation for a single frequency modulating function

Fig. 3.5

33

DATA TRANSMISSION

constant amplitude. Noise will sometimes change the frequency of a wave, but it is more likely to affect the amplitude. A receiver for a frequency modulated signal will at some stage deliberately limit the amplitude of the signal passing through it; any signal level above a certain fixed value is ignored. In this way a frequency modulated system reduces the effect of noise.

Frequency modulation is used more often for data transmission than amplitude modulation owing to this better noise performance. Although with, for example, binary data as the modulating function, the spectrum will be considerably more complicated than the single frequency discussed above. As with amplitude modulation there is a carrier frequency ω_1 with upper and lower sidebands.

3.4 Phase modulation

In this method the message signal y_2 is superimposed on to the phase angle of the carrier signal. Phase modulation is really a form of frequency modulation since it changes the angle. In practice it is difficult to detect small changes in phase, consequently the method is not normally used for analog signals such as speech and music. The method is used, however, to transmit binary data.

Let the carrier signal be again, for simplicity,

$$y_1 = A_1 \cos(\omega_1 t).$$

Consider changing the phase by adding to it another complete wave

$$y_2 = A_2 \cos(\omega_2 t).$$

Then

$$y = A_1 \cos(\omega_1 t + y_2)$$
$$y = A_1 \cos(\omega_1 t + A_2 \cos(\omega_2 t)),$$

where A_2 is the maximum change in phase. Since a phase change of $+\pi$ gives the same result as one of $-\pi$ the maximum values for A_2 are $\pm \pi$.

The factor A_2 is called the *modulation index*. As with frequency modulation, the function y involves Bessel Functions.
It can be shown to be

$$y = A_1 J_0(A_2) \cos(\omega_1 t)$$
$$+ A_1 J_1(A_2) [\cos((\omega_1 + \omega_2) t) - \cos((\omega_1 - \omega_2) t)]$$

MODULATION

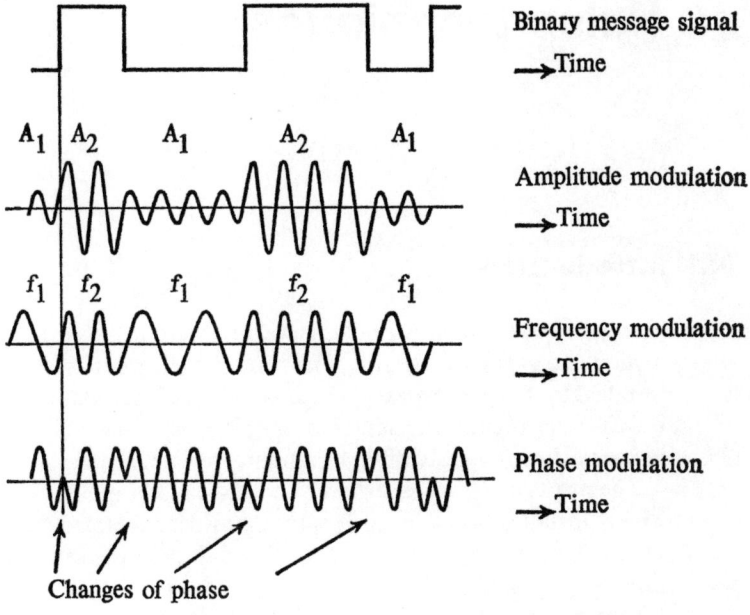

Fig. 3.6

$$- A_1 J_2 (A_2) [\cos((\omega_1 + 2\omega_2) t) - \cos((\omega_1 - 2\omega_2) t)]$$
$$+ A_2 J_3 (A_2) [\cos((\omega_1 + 3\omega_2) t) - \cos((\omega_1 - 3\omega_2) t)]...\text{etc.}$$

The function contains not only the frequency ω_1, but an infinity of other frequencies $\omega_1 \pm \omega_2$, $\omega_1 \pm 2\omega_2$, $\omega_1 \pm 3\omega_2$ ... etc, the amplitudes of each being given by a Bessel Function. Generally phase modulation uses a smaller bandwidth of frequencies than frequency modulation because of its smaller (restricted) modulation index.

Shown in Fig. 3.6 is an example of binary data as a message used to amplitude, frequency and phase modulate a carrier signal.

4 Pulse techniques

4.1 Introduction

The world's telephone services, generally, were installed to transmit analog messages or signals. The world's telegraph services were installed to transmit messages in a digital form. There has been a great increase in digital transmission requirements since about 1960, particularly on-line and off-line computer terminal equipment.

There are many cases, where, because the equipment is already widely installed, the telephone service has been adapted to transmit digital data. However, to transmit digital data at the same information rate as analog data does, in practice, require a wider bandwidth; or put alternatively, for a given line or channel the information rate must be smaller for a digital signal than for an equivalent analog signal.

Since the first telegraph transmission in the 1830's the capacity for frequencies of the world's transmission channels has been increasing exponentially. Up to about 1960 it could be said that bandwidth was at a premium; considerable effort was spent in devising methods of 'packing' messages so as to reduce their frequency range and yet not lose the information. However, with the gradual introduction of waveguides operating at the extreme radio frequencies of 10^{10} to 10^{11} Hz, and also with the development of the optical waveguide operating at frequencies of the order of 10^{14} Hz the bandwidth problem is becoming less acute. The scale of these improvements is illustrated by the fact that in principle one optical fibre cable could carry at the same time the whole of the world's current communication channels. There is thus no foreseeable limit to the bandwidths that will become available to carry messages.

There are two main forms of data, analog and digital. With no bandwidth restriction there is no doubt that coding an analog message into a sequence of digits, and transmitting it as a digital signal, has certain important advantages over analog transmission. One is that in the presence of noise (which is the case for all practical situations) a digital signal can be recognised with more certainty than an analog signal. Owing to signal losses in a channel any message must be amplified at many intermediate points of its transmission, the more so the higher the transmission frequency. In an analog signal every amplification stage increases the noise by the same proportion; the noise will accumulate. With a digital signal the same losses and therefore amplification requirement will apply, only this time at each amplification stage the digits or pulses can be reformed (or regenerated or reshaped) since the shape of each pulse will, in general, be known. Of course pulses will be lost from or falsely inserted into a digital message, but there are methods for detecting the presence of such errors, and methods for correcting them in certain forms of digital transmission.

The process of converting a message, whether it be in analog or digital form, into a sequence of pulses for transmission is called coding.

There are two main categories of pulse techniques. One is where some characteristic of the pulse is modified with the message; the second is where the message is converted into a sequence of pulses by some appropriate form of coding.

4.2 Pulse amplitude modulation (p.a.m.)

A sequence of pulses is transmitted, each one having an amplitude that contains some aspect of the message. The example shown in Fig. 4.1 is an analog signal directly sampled. Regular, equal time intervals are fixed, called the sampling interval; at these points the signal amplitudes are measured and pulses generated whose heights (or amplitudes) are equal to the signal amplitudes at those points. The inverse of the sampling interval is called the sampling frequency. The problem is: just how many samples of a particular signal need be taken for it to be adequately represented?

Consider a set of points representing a sampled signal (see Fig. 4.2). The number of different waves that can be fitted through the points is infinite (see Fig. 4.3). If, however, the time between samples is b, then there is only *one* wave with frequencies less than $1/2b$ which

DATA TRANSMISSION

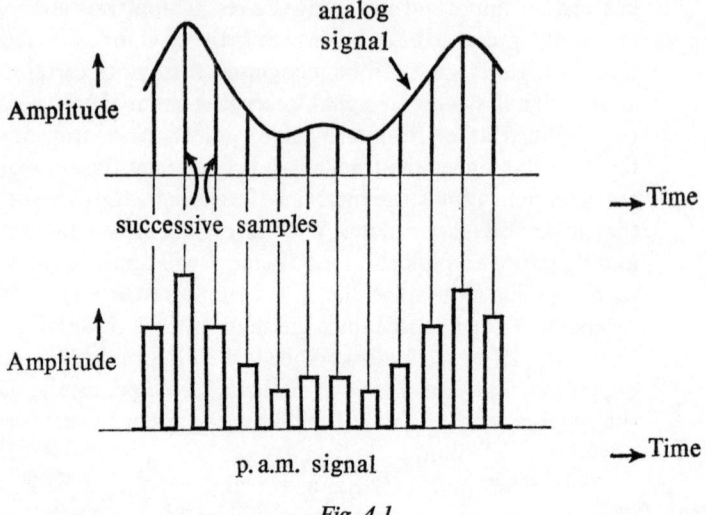

Fig. 4.1

can be fitted through the points. Conversely, it can be shown that if a signal contains frequencies up to but not including W, then sampling at $2W$ samples per second, or more, is sufficient to be able to reproduce the wave uniquely. This means that a single frequency must be sampled at least three times in one wavelength. That the signal sampled $2W$ times per second must not contain a frequency *equal* to W is illustrated in Fig. 4.4, which shows a wave of single frequency W, sampled precisely $2W$ times per second—the result always being zero. Therefore a signal containing frequencies up to but not including W Hz sampled $2W$ times per second is sampled adequately for the

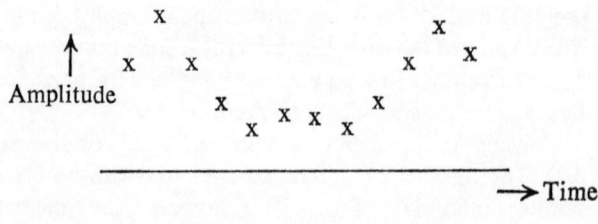

Fig. 4.2

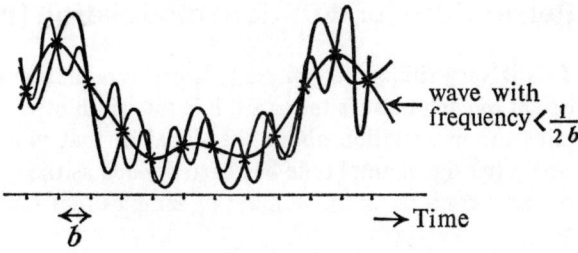

Fig. 4.3

high frequency range, and more than adequately for the lower frequencies.

W is sometimes called the folding frequency, as any frequency above W is 'folded' back into the range 0 to W. In practice it is usual to sample well above $2W$ to counteract these types of errors. A baseband analog signal limited so that it contains no frequencies *equal to* or greater than 4000 Hz will be correctly sampled 8000 times per second (or more).

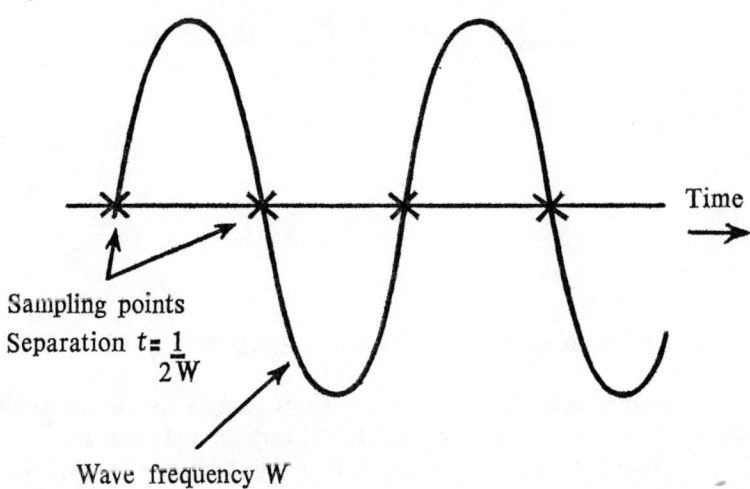

Fig. 4.4

DATA TRANSMISSION

4.3 Pulse width (or duration) modulation (p.w.m.)

This is very similar to pulse amplitude modulation except that the height of each pulse is the same, it is the width of the pulse that contains the information about the signal at that point, see Fig. 4.5, where the signal amplitude has been defined as the information. The remarks concerning the number of samples that must be taken still hold.

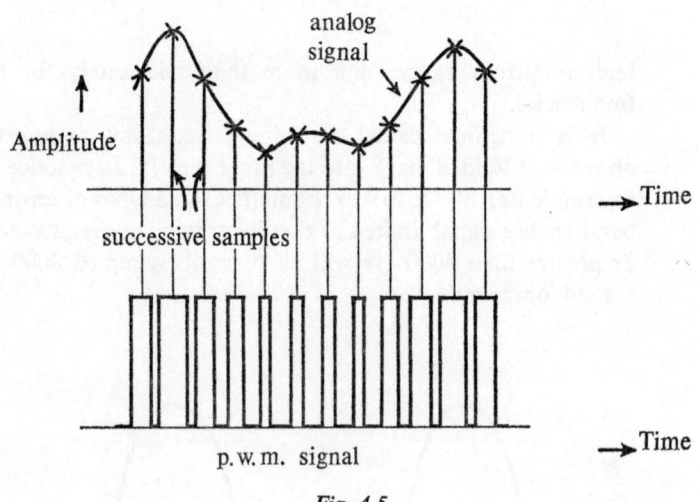

Fig. 4.5

4.4 Pulse position modulation (p.p.m.)

In this method the height and width of the pulses are constant, but their position in the output signal varies about some mean position according to the signal at that sampling point. (See Fig. 4.6.) The remarks concerning the number of samples that must be taken still hold.

PULSE TECHNIQUES

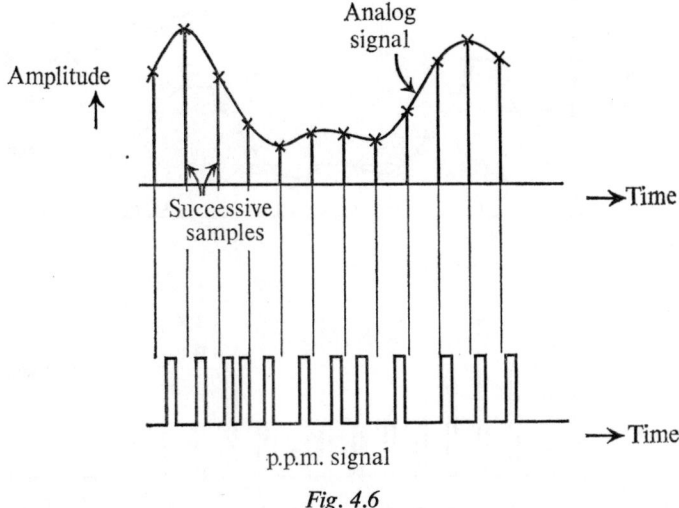

Fig. 4.6

4.5 Pulse code modulation (p.c.m.)

The above three examples, p.a.m., p.w.m. and p.p.m., still effectively represent an analog signal. It is more usual to find that the sample information, such as amplitude, at each point is coded in some way; for example as a binary number. This is called pulse code modulation.

To code the analog amplitude in binary the axis is divided up into a discrete number of equal intervals; 32 levels for a 5-bit number, 64 levels for a 6-bit number, etc. At each sampling point the amplitude is measured and the level nearest that value taken as the approximation to the amplitude. Figure 4.7 shows the stages in obtaining a binary output as the p.c.m. with 8 levels. Therefore there will always be an error, the maximum value of which is $\frac{1}{2}$ × separation in levels (see Fig. 4.8). Most current techniques transmit the binary number serially as shown in Fig. 4.7. The bit rate on the line must be much higher than the rate at which the analog signal is being sampled. For example, sampling an analog signal 8000 times per second, coding each sample as a 10-bit number and transmitting it serially results in a bit rate of 10 × 8000 = 80,000 per second. For this reason most techniques incorporating p.c.m. are used on high speed (wide bandwidth) lines. There is a trend, however, towards multilevel trans-

DATA TRANSMISSION

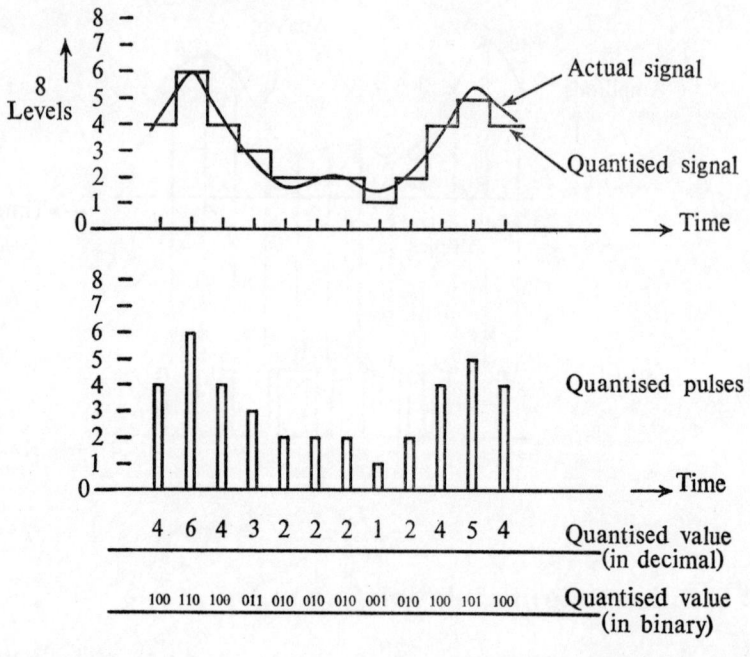

Fig. 4.7

mission; for example, a data transmission system working in ternary; i.e., a three level system. This refers to the permissible line signal states. The manner in which transmitter and receiver use these states is a first-stage design choice. The next major step forward will probably be a true multilevel transmission system combining the advantages of binary transmission and analog transmission.

The sampled amplitude can be, and is coded in many different ways, e.g. as a decimal number. One method divides the amplitude axis into a discrete number of unequal intervals, the steps chosen to be on a logarithmic scale. The idea here is that analog noise which is usually at a fairly low level is then, when on its own, always sampled

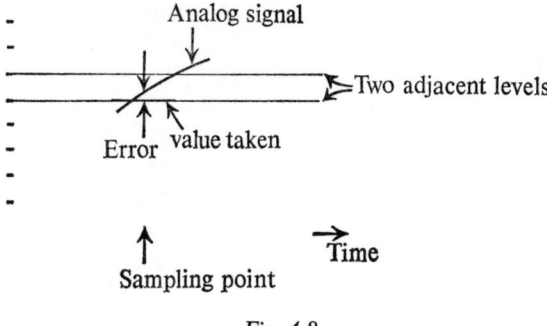

Fig. 4.8

and coded as zero (or near zero), while higher levels of more meaningful signal have correspondingly more weighting.

Pulse code modulation is a form of quantising in two dimensions. One is the time axis (by sampling) the second is by only permitting a finite number of amplitude levels. Consider the case of each sample being represented by a 10-bit number (i.e. 1024 levels); this is fairly typical of analog to digital converting devices. If the output from the device is in serial form on one line, and if the analog signal input is limited to frequencies less than 4000 Hz, then with a sampling frequency of 8000 the number of bits per second output is 10 × 8000. This illustrates just one of the problems of signal analysis by digital computer, namely the large amount of data. One minute of real time signal sampled 8000 times per second is 60 × 8000 samples; if it is going to be used as data in a 40-bit word computer, and there are 10 bits per sample then there are 60 × 8000/4 = 120,000 computer words packed with 480,000 numbers.

Pulse code modulation, then, never permits the analog signal to be exactly recovered, since the coding allocates a finite number of codes (e.g. binary numbers) for an infinite variation in analog signal level. For example, 7 bits could be allocated for a signal level which may be anything in the range 0 to 6 volts. In this case each sample is accurate to $\frac{1}{2}$ part in 2^7 (or 1 in 256). Reforming the analog signal from this digitised and coded signal reproduces this inaccuracy as a form of noise. It is called *quantisation noise*. Figure 4.9 shows how quantisation noise varies with quantisation levels.

DATA TRANSMISSION

Number of Sampling levels	Quantisation Noise dB	Quantisation Noise (*relative levels*)
32	− 36	1 in 64
64	− 42	1 in 128
128	− 48	1 in 256
256	− 54	1 in 512
512	− 60	1 in 1024
1024	− 66	1 in 2048

Fig. 4.9

4.6 Modulation with orthogonal digital carriers

In Chapter 3 the classical concepts of amplitude modulation (a.m.) frequency modulation (f.m.) and phase modulation (p.m.) were discussed. A general description of the process is that the information to be transmitted is combined in some way with the special orthogonal functions Sine and Cosine as the carriers. There is, in principle, no reason why other orthogonal functions should not be used for this process. There is some indication that a successful transmission system can use Walsh or Haar functions: these, though, are digital functions, and used as carriers require a new definition of the concept of frequency. At the time of writing rapid developments are being made in this field—see references 4.1, 4.2 and 4.3 for further reading.

5 Noise, error detection and correction

5.1 Noise

In a communication system there is noise present in all practical situations. Noise is defined as any signal that interferes with the message being sent. It can be another unwanted message or a random fluctuation in level. Examples are: the sound from aircraft engines and the interference of television by electrical apparatus. Certain weather conditions can cause two broadcasting stations to overlap in the frequency domain; this is heard on the domestic radio as the two stations being received simultaneously.

The effect of noise depends not only on the type and level of the noise, but on the actual message being transmitted. For example, if the message signal is a sequence of binary digits representing a computer program then every bit is vital. But if the message is English text then a certain proportion of the characters can be wrong and the message still be received and understood. It is sometimes said that operator mistakes are a form of noise; for example, mistyping a message text, on the grounds that the effect—which is the corruption of the message signal—is the same. More usually noise is random interference.

White or Gaussian Noise

This can be heard as a background hiss on the domestic radio. In electronic circuits it originates from many sources, one being the slightly random motion of electrons through a conductor; the overall current motion has added to it another, usually low level, random effect which is temperature dependent and so it is sometimes called thermal noise. The amplitude of the noise follows a random fluctuation about a mean value, the distribution of amplitudes being Normal (or Gaussian). In some experimental laboratories work is carried out on channels with hundreds of voiced signals multiplexed together.

DATA TRANSMISSION

The combined signal output from this channel is very much like Normal (or Gaussian) noise; this fact is used to simulate the signal for test purposes. Figure 5.1 illustrates a typical Gaussian pattern. There are noise sources giving random signals that obey other statistical laws; for example, with amplitudes following a Poisson distribution.

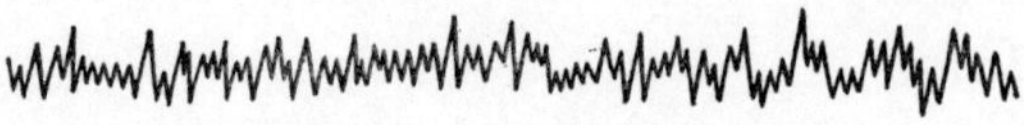

Fig. 5.1

Impulsive noise

This is random in the sense that it is not predictable. It produces large amplitude levels for short periods of time that can easily saturate a message signal. There are many sources, the most common being the making/breaking of electrical contacts; for example, electric light switches, switch banks in telephone exchanges, automobile ignition systems, lightning. The effect of this form of noise is greatest on high-speed data. A typical impulsive noise burst could last for 10 msecs which to a listener on a telephone line would be a sharp click. But on a data line carrying digits at, for example, mega bit (10^6) rate it could destroy 10,000 digits.

Cross talk

If two lines carrying electrical signals are run near to one another then by inductive and capacitive coupling the signal of one appears on the other and vice versa. This is called cross talk. Proper screening, matching and earthing can reduce it to negligible proportions but never entirely eliminate it.

Intermodulation distortion

In the discussion of amplitude modulation it was shown how two signals with differing frequencies were deliberately chosen to interact with one another to produce a different set of frequencies. On any line carrying multiplexed signals there is the risk of the same effect happening, accidentally; that is the interaction of two or more signals to produce a third different unwanted signal. This is called intermodulation, and when the unwanted signal interferes with a

NOISE, ERROR DETECTION AND CORRECTION

transmitted message it is called intermodulation distortion. For example, a data transmission line could have multiplexed together single frequency control signals as well as two or more data messages. If the multiplexing is not very carefully arranged and the frequencies correctly chosen it could easily happen that one of the single frequencies could modulate one data stream on to another.

Distortion

All channels have a set of physical characteristics (see below); the effect on a signal using the channel is that it is transformed according to the characteristics. This is called distortion. In practice it means that a received signal will be different in some degree from the transmitted signal. On electrical channels typical forms of distortion are:

(a) Frequency distortion: different frequencies are attenuated by different factors, so a square wave which consists of a large range of frequencies could be deformed as shown in Fig. 5.2.

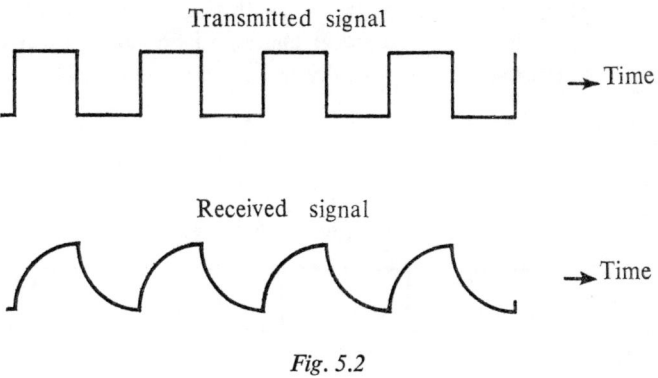

Fig. 5.2

(b) Phase distortion: different frequencies have different delay times. The possible effect on a square wave is shown in Fig. 5.3.

(c) Frequency change: a signal transmitted with frequency ω_1 will be received as one with frequency ω_2.

DATA TRANSMISSION

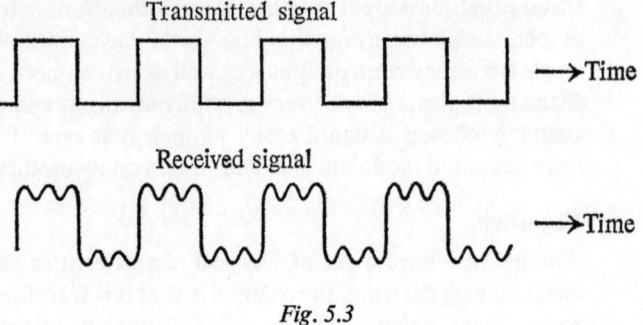

Fig. 5.3

(d) Amplitude distortion: different amplitudes are attenuated by different factors. A device called a compandor deliberately uses this as a technique to improve the signal/noise ratio in speech transmission.

(e) Bias distortion: On a transmission line with repeaters the reconstruction or reshaping of a pulse can result in its being made wider or narrower. Figure 5.4 illustrates a possible result. On channels where the distortion characteristics are known compensation can be used which partially and sometimes completely eliminates it.

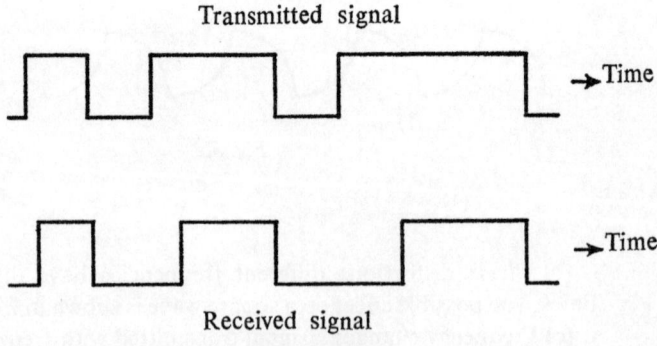

Fig. 5.4

NOISE, ERROR DETECTION AND CORRECTION

5.2 Error detection and correction

Consider the situation in which two possible messages are to be transmitted. They may or may not be complicated but there must be only two. For example:

(i) Restock 100 items part number 1000 model 100
(ii) Restock 200 items part number 1001 model 500

Then, provided both transmitter and receiver knew beforehand the contents of the two messages communication could take place in the following three ways:

(a) A simple binary transmission could indicate either of the messages.
  Reception of binary 0 would indicate message (i)
  Reception of binary 1 would indicate message (ii)

In practice there would be errors; for example, owing to noise. An error would result in the wrong message being received, but could be guarded against by transmitting the binary indicator twice.

(b) So message (i) would be shown by two binary 0's and message (ii) would be shown by two binary 1's.

If either 01 or 10 is received then an error has occurred; which of the two messages is indicated is not known. It might be thought that in this case it would be a better strategy to transmit the single binary digit—situation (a). But, of course, this could lead to the wrong message being received with no knowledge of the error. There is a further possibility of two errors occurring changing 00 into 11 or vice versa.

(c) So the transmitted signal could be made more complicated, using three binary digits for the transmitted indicator.
  If 000 is received then the message is (i)
  If 111 is received then the message is (ii)
  If 001, 010, or 100 is received then there is an error, but the message is still (i) since there are more 0's.
  If 110, 101, or 011 is received then there is an error but the message is (ii) since there are more 1s.

The above three situations are examples of three ways of coding the transmission of two possible messages. They illustrate how by making the code more complicated the effect of noise is reduced—

since the probability of three successive errors is usually much less than the probability of one error. Situation (b) shows a method of *detecting* single errors. Situation (c) shows a method of *detecting* and *correcting* single errors.

Situation (c) is interesting because it illustrates that by making the code more complicated (using three bits) to counteract errors has, in a sense, reduced the efficiency. Three binary digits would have permitted the communication of eight different possible messages, but then if any one bit had been wrong the message would have been corrupted (situation (a) in fact). Using three binary digits to transmit eight possible messages can give no detection or correction of errors but to transmit two possible messages can correct single errors and detect single or double errors. If the chance of three successive errors is high then a code could be made from four bits, five bits, six bits, etc, every time error detection/correction (i.e. complex coding) has to be traded against efficiency (the number of different possible messages required).

In the example chosen there are 44 characters (including spaces); the ultimate coding would have been to allow 5 bits for each character. Then there is a transmission of 44×5 bits. The choice would be either two possible messages transmitted—communication being by 220 binary 0's or 1's with extremely good error correction—or 2^{220} different possible messages. In reality the situation is chosen to be somewhere between the two.

Situation (b) and particularly situation (c) illustrate *redundancy*. Redundancy is the transmission of more information than is strictly necessary for communication to take place in order to guard against corruption (both by noise and operator mistakes). See section 1.3.

Redundancy is to be seen in the natural languages where with a given group of, say, three letters there is a high probability that one particular letter will follow. In the English language the probability that 'N' follows 'TIO' is very high; the 'N' is almost redundant. But had the 'T' or 'I' or 'O' been corrupted then the 'N' is important confirmatory evidence. This is to be seen with words also; in English the probability that 'that' follows 'have a high probability' is high. Letters, words, sentences confirm one another to a considerable extent. English language could be compacted by a third and still be meaningful, but then corruption would take on significant proportions. See references 1.2 and 1.3.

NOISE, ERROR DETECTION AND CORRECTION

Natural languages evolved redundancy to counteract corruption by noise. This is because communication took place by speech and also by writing. The acoustic path (i.e. channel) over which the sounds of speech travel is highly susceptible to interference by noise. The principles of data communication are the same: if the channel is particularly noise free then simple error detection is probably sufficient; if the channel is noisy then more redundancy must be built into the coding so that the errors are counteracted.

One method of making sure that the transmitted signal is the one that arrives is for the receiver to retransmit the received message back to transmitter which then checks the two messages. It is not very efficient because the receiver always retransmits even if the message has been received without errors. Nevertheless where time is not important it is a method used—usually on short, full or half-duplex lines (see section 6.1).

Parity checks

In any particular system the complete set of characters is coded; for example, the 7 bits of the ASCII code, and a further bit added so that

Information	Parity digit	
0 0 0 0	1	sum of binary 1's is odd
0 0 0 1	0	
0 0 1 0	0	
0 0 1 1	1	
0 1 0 0	0	
0 1 0 1	1	
0 1 1 0	1	
0 1 1 1	0	
1 0 0 0	0	
1 0 0 1	1	
1 0 1 0	1	
1 0 1 1	0	
1 1 0 0	1	
1 1 0 1	0	
1 1 1 0	0	
1 1 1 1	1	

Fig. 5.5

DATA TRANSMISSION

the sum of binary 1's in any single character is either an odd number (*odd parity*) or an even number (*even parity*). Both systems are in common use. Figure 5.5 illustrates odd parity with a 4-bit code. It detects single digit errors, that is, a 1 changed to a 0 or vice versa. Five-hole paper tape could have odd parity for each character, 4 bits for the information, one bit for the parity digit (it must be odd to allow for the rub out, erase or all holes character), and eight-hole paper tape could have even parity for the same reason.

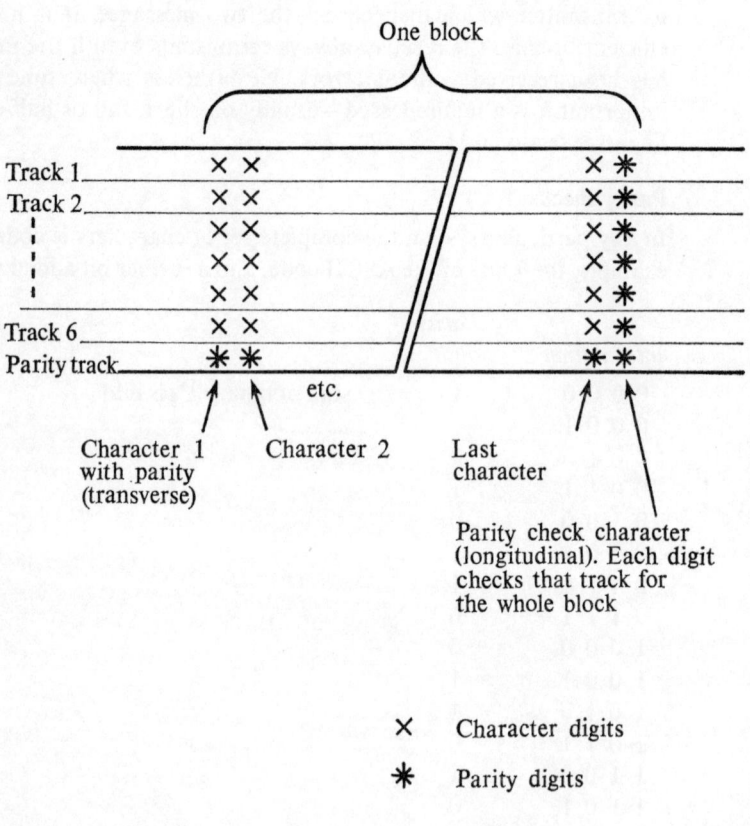

Fig. 5.6

NOISE, ERROR DETECTION AND CORRECTION

Often characters are gathered into blocks, for example of 100 characters, and a further special character added as a block parity check. This is used on magnetic tape systems; the information parity is called a transverse parity check and the block parity is called a longitudinal parity. See Fig. 5.6.

The longitudinal and transverse parity are useful double checks for errors. Often the hardware associated with this system will stop on detecting a parity failure, indicate a parity fault by lights and wait for operator intervention. In principle it can be used to correct single digit errors, because if the transverse parity has failed for example on character 3 and the longitudinal parity failed on Track 2 then the second digit of character 3 must be wrong and can be corrected accordingly. The problem is that the machine must store the parity failure information for at least one block.

Error correction

The deliberate addition of redundancy (digits) to a message in order to correct for possible errors in called Forward Error Correction, as opposed to error detection with a request for retransmission. The previous paragraph describes an example of correcting single-digit errors.

It is generally true that error-correcting codes, to be efficient, must be complicated. Often there are as many redundant digits as exist in the original message—see reference 5.1 (the parity digits are the redundancy). Error correction is useful on *simplex* circuits (for example, broadcast) and in situations where retransmission is not practical—for example, on data which is to be stored on magnetic tapes for any length of time. It has been found that for data transmission on full- and half-duplex circuits the best compromise is error detection with an automatic request for retransmission—see Chapters 10 and 11.

The single-digit parity check informs the receiver of a single error; it does not say which digit is wrong. The further block parity is often difficult to implement. A more common improvement is to add to the character another parity digit, making two in all. For example, in a 6-bit character parity digit 1 could check information bits 1, 2 and 3 and parity digit 2 could check bits 2, 3 and 4: this gives a total of six bits, and a system for detecting double (and single) errors. Figure 5.7 illustrates this with even parity. Closer examination of this

DATA TRANSMISSION

double-parity system shows that it will not only detect single and double errors but goes part way to correcting single errors. Suppose parity check 1 fails but parity check 2 passes, then the information digit 1 must be in error. Suppose parity check 1 passes but parity check 2 fails, then digit 4 must be in error. But if both parity checks fail then either digits 2 or 3 could be wrong.

Information bits 1 2 3 4	Parity 1 (even, on digits 1, 2 and 3)	Parity 2 (even, on digits 2, 3 and 4)	Total Character
0 0 0 0	0	0	000000
0 0 0 1	0	1	000101
0 0 1 0	1	1	001011
0 0 1 1	1	0	001110
0 1 0 0	1	1	010011
0 1 0 1	1	0	010110
0 1 1 0	0	0	011000
0 1 1 1	0	1	011101
1 0 0 0	1	0	100010
1 0 0 1	1	1	100111
1 0 1 0	0	1	101001
1 0 1 1	0	0	101100
1 1 0 0	0	1	110001
1 1 0 1	0	0	110100
1 1 1 0	1	0	111010
1 1 1 1	1	1	111111

Fig. 5.7

R. W. Hamming—see reference 5.2—used multiple parity checks for detecting errors and correcting some of them. An example is a code called the 4 out of 7 code which uses four bits for information and three for parity checks. It will either detect two (or one) errors or will locate (that is, correct) single errors. Parity digit 1 checks digits 1, 2 and 3, parity digit 2 checks digits 1, 2 and 4 and parity digit 3 checks digits 1, 3 and 4 (see Fig. 5.8, which illustrates even parity).

NOISE, ERROR DETECTION AND CORRECTION

Information digit no.	Parity 1. Checks 1,2,3	Parity 2. Checks 1, 2, 4	Parity 3. Checks 1,3,4	Total Character
1 2 3 4				
0 0 0 0	0	0	0	0000000
0 0 0 1	0	1	1	0001011
0 0 1 0	1	0	1	0010101
0 0 1 1	1	1	0	0011110
0 1 0 0	1	1	0	0100110
0 1 0 1	1	0	1	0101101
0 1 1 0	0	1	1	0110011
0 1 1 1	0	0	0	0111000
1 0 0 0	1	1	1	1000111
1 0 0 1	1	0	0	1001100
1 0 1 0	0	1	0	1010010
1 0 1 1	0	0	1	1011001
1 1 0 0	0	0	1	1100001
1 1 0 1	0	1	0	1101010
1 1 1 0	1	0	0	1110100
1 1 1 1	1	1	1	1111111

Fig. 5.8

The following combinations of parity and information digits

Information digits	Parity digits	Total digits per character
1	2	3
4	3	7
11	4	15
26	5	31

give codes which detect one or two errors and correct one error. The second is the four out of seven code. More complicated coding is needed to correct two or more errors, and/or detect three or more.

The problem facing the designer of a data transmission system is that errors often occur in groups or bursts widely separated in time. For much of the transmission time there will be no errors so correcting codes are wasteful, but when they do occur it could be a multiple error; this is typically the effect of impulsive noise.

DATA TRANSMISSION

For further reading on error correction see ref. 5.1. Particularly interesting is the chapter on Recurrent Codes which deals with the burst-error-correcting codes of Hagelbarger (see also refs. 5.3 and 5.4).

There have been many measurements made of the transmission errors found in practical situations. A typical average error rate on 50 baud Telex lines is 1 in 10^5; see ref. 5.5.

6 Parallel and serial transmission

6.1 Definitions

A channel is a path for the transmission of information between two or more points. Other names that are used for it are circuit, facility, line, path, wire. A channel is further classified as
(a) *Simplex* if it carries information in only one direction. For example, a line used to link a fluid level measuring instrument in a tank to a computer.
(b) *Half-duplex* if it can carry information in either direction but not simultaneously. For example, the air-powered tubes some pre-1940's department stores used to convey bills and money to and from the cashier and customer.
(c) *Full-duplex* if it can carry information in either direction at the same time. For example, the telephone line. Simple electrical circuits for the above are illustrated in Fig. 6.1.

These definitions are applied by most of the world's computer and telecommunications engineers and will be used in this book. Unfortunately Europe's telecommunications engineers use the term simplex for a half-duplex line, and half-duplex for a full-duplex line that is limited by its terminal equipment to half-duplex working. Thus in Europe computer and telecommunications engineers often mean different things when talking of simplex and half-duplex lines.

6.2 Parallel transmission

Each element of a character (or code) is transmitted along its own channel, so that the total character is transmitted at the same instant. For example, a 5-bit character needs 5 channels. Because it uses a number of channels it is expensive but is essential in some situations. A device, the British Standard Interface (see section 11.5), uses parallel transmission particularly suitable for computer use.

DATA TRANSMISSION

Tx = Transmitter

Rx = Receiver

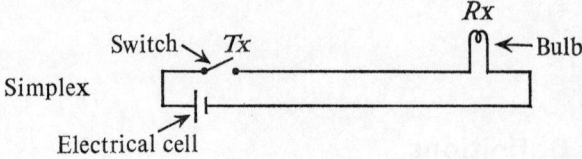

Simplex

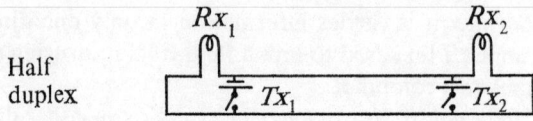

Half duplex

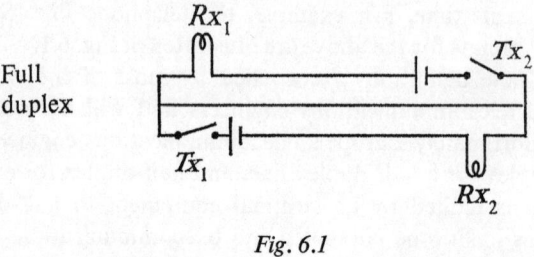

Full duplex

Fig. 6.1

Often the terminal equipment is cheaper with parallel transmission. It is commonly used for short line lengths where the user has control over both the transmitter and receiver. Some machines designed for parallel operation use separate channels not on separate lines but by multiplexing within one line. Frequency division multiplexing is often the method used.

PARALLEL AND SERIAL TRANSMISSION

6.3 Serial transmission—asynchronous

The information in the form of characters (or codes) is transmitted one element at a time (i.e. serially) along one channel. For example, a terminal teletypewriter on-line to a computer could code each character into 8 bits. In serial transmission they would be transmitted one bit at a time. Thus the ASCII code for the letter S is 01010011 and it would be transmitted as shown in Fig. 6.2 along one channel.

ASCII Code for the letter S

Fig. 6.2

For the receiving equipment to decode a character correctly it must 'know' when to expect it, or where to find the information. It must be able to lock into the transmitted signal information. This is called synchronisation, and is a timing problem.

Asynchronous transmission is the form used by most electro-mechanical serial devices such as teleprinters, teletypewriters, etc. Every character consists of three parts a *start* element, the information, and a *stop* element. Normally the information is sent as a fixed number of bits, usually 5 or 8. These three parts are transmitted serially along one channel, each part occupying a certain length of time. It is usual for the start element to be the same length or time width as 1 information unit. The *stop* element is either 1, 1½ or 2 times the length of the information unit; it is machine dependent, a length of time that allows mechanical rods and levers to be reset. Thus a serial asynchronous character could appear as shown in Fig. 6.3.

The *start* and *stop* elements are always of opposite polarity. Typical operating situations on the line are 0 to 6 volts; -6 to $+6$ volts; -6 to 0 volts; -60 to $+60$ volts. In telegraphy the terms MARK and SPACE are used. The stop or reset state is always a mark, the start a space. When data is not being sent the line is at the same state as the stop element. The mark is usually a 1, a hole in paper tape, a

59

DATA TRANSMISSION

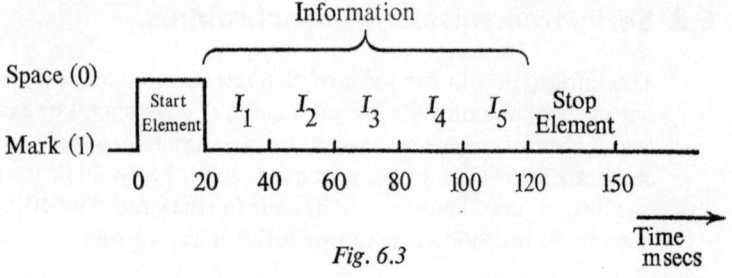

Fig. 6.3

current, a dot, etc; the space is the converse of the mark. The beginning of a character is always defined as the transition from *stop* to *start*, which must therefore be a change in polarity. As soon as the stop/start transition is detected then some timing device or clock samples the signal as it is being received on the line. Thus the receiver must have a complete description of the character; that is, it must know how long in time an element lasts, how many information units there are and how long the start element is.

The advantages of asynchronous serial transmission are: that the characters can be generated easily by electromechanical devices, and can be easily used to drive such devices, for example teleprinters; that the character is itself complete, it contains its own synchronising information, namely the start and stop elements; that it does not matter how regular or irregular in time the characters are transmitted or received.

The disadvantages are: that a considerable section of the character is not bearing message information, it is thus not very efficient; that synchronisation is dependent upon recognition of the stop/start transition, which is fairly easily missed or falsely discovered in the presence of noise; an electrical path always distorts the signal, asynchronous serial transmission is sensitive to distortion; that the speed must be limited because of the sensitivity to distortion.

6.4 Serial transmission—synchronous

A serial signal stream is transmitted along one channel—as before—except that in this case there are no stop and start elements. Synchronisation is carried out by gathering the data into blocks, for

PARALLEL AND SERIAL TRANSMISSION

example 100 characters, and prefixing every block with a unique code which will be recognised by the receiver. Detection of the unique code causes the receiver to synchronise with the signal for the whole of that block.

Synchronising information must always be sent with the message; this information can be generated either by the transmitter, in which case the receiver must have prior knowledge about it, or it can be generated by a source common to both transmitter and receiver. The receiver upon detection of the synchronising code uses some timing mechanism; for example, a square wave clock, to pick the information from the message signal. The receiver must still know precisely the format of the incoming signal, and it must follow this format very closely.

Any timing error in asynchronous transmission causes the loss of one character; it could lose a whole block in synchronous transmission.

The advantages of the synchronous system are: that it is more efficient since the proportion of message to synchronising information is greater; that the unique synchronising code can be made complicated (e.g. the characters SYN SYN SYN), this gives a high probability that synchronisation will take place; that it is not so sensitive to distortion and can operate at higher speed; that a common timing mechanism (clock) can be used for transmitter and receiver.

The disadvantages are: that if a synchronising error does occur then a whole block is lost; that the characters must be sent in blocks and not as they become available; that the equipment tends to be more expensive (at the time of writing).

The table in Fig. 6.4 shows the comparative operating speeds and equipment for each system.

Speed	Asynchronous	Synchronous
LOW 0 to 300 Baud	Electromechanical devices teletypes, teleprinters	Rarely used
MEDIUM 300 to 5000 Baud	Unbuffered terminals such as papertape readers	Buffered terminals such as line printers
HIGH 5000 → Baud	Rarely used	Computer→Computer

Baud: Unit of signalling speed (after Emile Baudot). In an equal length code it is the number of discrete signal events per second. If each signal event is one bit, then Baud speed equals bits per second.

Fig. 6.4

6.5 Modems

The word *modem* is a contracted form of the words modulator/demodulator.

The concept of a transmission system as a transmitter, channel(s) and a receiver is logically complete. That is, the transmitter can be thought of as applying a signal to a channel(s) and the receiver as collecting that signal. However, in practice there are many problems in setting up such a system, one is matching the transmitter and receiver to the channel that is used.

The idea of analysing any signal, analog or digital into a set of frequency components has already been discussed in Chapter 1. The way in which a *modem* operates is best thought of in terms of these frequency components, that is, it is to be pictured working in the frequency domain rather than the time domain.

Modems are not always used. The world's telegraph systems were devised to transmit digital signals; the transmitter and receiver were designed to match the channel being used. In the early days this was a pair(s) of wires—called open pair(s)—carried on poles. Today telegraph circuits are still used, and often a form of modulation/demodulation is used to enable the line to be more efficiently employed.

So the modem is to be thought of as essentially a matching device working in the frequency domain. The most common example is the utilisation of the telephone communication network to transmit digital information. Dialling a number sets up connections in ex-

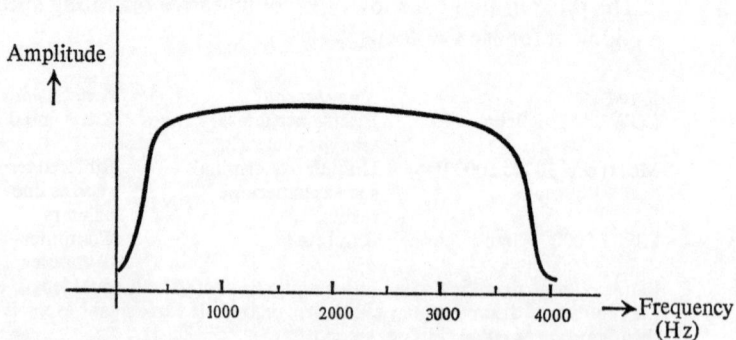

Fig. 6.5. Typical characteristics of a telephone channel. Frequencies below approximately 100 and above 3,500 are heavily attenuated.

PARALLEL AND SERIAL TRANSMISSION

changes to the subscriber with that number. In this case a full duplex channel is established. The only problem is that this channel was designed to carry the human voice, which is an analog signal having a frequency range of approximately 300 Hz to 4000 Hz. The channel characteristics are pictured in Fig. 6.5.

How, then, can such a channel be made to carry a digital signal (e.g. see Fig. 6.6) with frequency components from zero (the d.c. term) upwards? A modem is a device which can transform in the frequency domain. It can convert the baseband digital signal to one lying in the range 300 to 4000 Hz. Modems can use any of the modulation processes described in Chapter 3. Figure 3.6 illustrates a binary signal modulated in three ways—amplitude, frequency and phase—all possible modem actions.

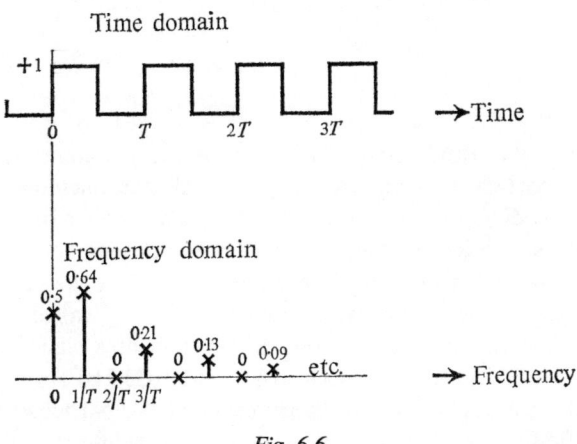

Fig. 6.6

It is interesting that as soon as a modem mechanism is admitted to the system then a number of developments follow. One is to multiplex on one simplex channel with telephone characteristics a full duplex circuit, each channel having half the bandwidth of the whole. Another is to allocate on a similar simplex channel one half of the frequency band for data to be conveyed in one direction, and the other half of the band for control signals (associated with the data) to

DATA TRANSMISSION

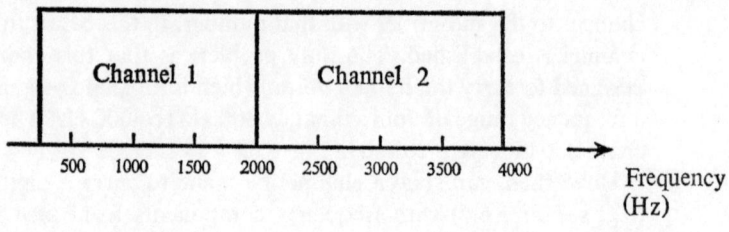

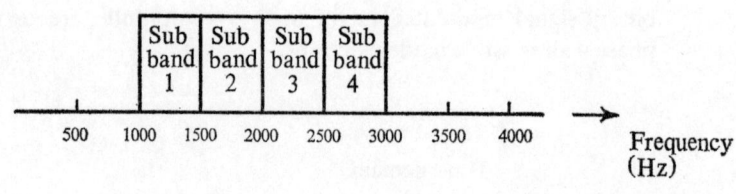

Fig. 6.7

travel in either direction (see the B.S.I. in section 11.5). Yet another is to take the same simplex channel with telephone characteristics and divide the frequency band into a number of sub-bands, using these sub-bands as a parallel transmission channel (see Fig. 6.7).

A further development of this is to reduce the sub-bands to single frequencies. This gives what is commonly called multitone transmission. For example, one Bell-system uses eight frequencies, 697, 770, 852, 941, 1209, 1336, 1477 and 1633 Hz. Any one character is transmitted in parallel by being coded as a particular combination of frequencies. One problem here is that intermodulation products should not produce allowable frequencies and so corrupt this character being transmitted (see Chapter 3), hence the apparently strange choice of frequencies.

7 Store-and-forward and circuit switched networks

7.1 Introduction

In most countries there are historically two types of communication networks, telegraph for digital character transmission and telephone for analog (voice). However, the division between the two types has become increasingly blurred, particularly since about 1960. Telephone lines now carry digital data, which together with developments such as satellite and waveguide communication makes the definition of, say, a telephone circuit particularly complicated.

Before 1960 communication with a computer was carried out by the user (programmer) personally acting at the control consol. The development of large multi-access machines with many terminals 'on-line' has relegated the computer user to a role of machine sharer. The communication system of users connected to large multi-access computers has become increasingly complicated. If it is regarded as a data transmission system then there are a number of comparisons to be made with a telephone network.

7.2 Circuit switching

Both the telephone and the telegraph (telex) networks use circuit switching. Dialling causes a particular combination of electrical contacts to be made which link the caller to the dialled subscriber. Thus established it is a fixed electrical path. A line on which such a call is taking place is said to be busy, and it is held completely for the duration of the call. The system is called *Through Circuit Switching*.

In data transmission the connect time is frequently a period of half an hour or more during which time the data is often transmitted in sporadic bursts. There are therefore two disadvantages of through circuit switching for data transmission; one is the inefficient use of

DATA TRANSMISSION

the connect time, the other is that a busy line cannot be accessed by another user. To partially overcome these disadvantages a system called *Fast (or Rapid) Circuit Switching* has been proposed. A call is broken down into a series of subcalls; each subcall is a burst preceded by a short dial up time (e.g. 100 msecs) and followed by a rapid disconnect. Different calls can therefore be interleaved on the same circuit, but although the efficiency is improved the total dial up and disconnect times can be large compared with the call time.

7.3 The store-and-forward concept

(a) Message switching

In this system the information to be transmitted is gathered together by the transmitter into a special format and stored ready for transmission at a suitable later time. A single message consists of three parts: (i) a header which is the address or destination of the message; (ii) the message; (iii) an end of message marker. Sometimes (iii) is omitted, but in this case the single message must be a fixed and agreed length. The system transmits information as a sequence of single messages.

There is no dial up, the message header always governs its destination.

Some disadvantages are that the system must be able to store information. At the time of writing store is fairly expensive; and large delays are possible. The system is frequently not flexible enough.

(b) Packet switching

This is an extension of message switching. The data stream, or message, is divided into a sequence of convenient pieces called packets. These packets are then transmitted as individual messages, one packet only being sent when it is assembled and ready to go. Packet switching has been suggested as a method for interfacing all types of peripherals.

One disadvantage is that the system is only really suited to digital data transmission, particularly that involving a computer.

Store and forward methods always increase transmission time owing to the added address, but a great advantage is in the control of errors. In this respect the method of handshake is common: a hand-

STORE-AND-FORWARD AND CIRCUIT SWITCHED NETWORKS

shake system involves the receiver acknowledging the safe receipt of the information before the next is transmitted (see section 11.5). The transmitter stores the transmitted message, only erasing it when this acknowledgement is made.

A comparison made between store and forward methods and circuit switching has indicated that packet switching gives a relatively better performance up to surprisingly long messages. References 7.1, 7.2, 7.3 and 7.4 contain useful information.

8 Multiplexing and multiplexers

8.1 Multiplexing

Multiplexing is the utilisation of a channel by dividing it into two or more channels. An example is the domestic radio. The signal input to a domestic radio set contains a multiplicity of transmissions from many stations. From the point of view of the receiver it is just one signal, but by suitably adjusting electrical circuits a particular broadcasting station required can be selected from this multiplexed signal.

There are many forms of multiplexing; examples are frequency division, time division, space division and p.c.m. multiplexing. Whenever a channel has a greater capacity for frequencies than the signal being transmitted along it then theoretically the wasted capacity can be used either to take another channel or to increase the amount of information in the existing one.

Frequency division multiplexing (f.d.m.). If it is required to transmit a number of baseband signals along one channel then either

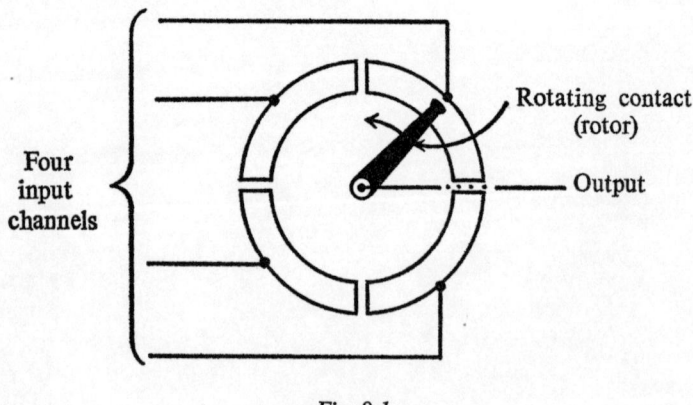

Fig. 8.1

MULTIPLEXING AND MULTIPLEXERS

amplitude, frequency or phase modulation can be used to multiplex the signals together in the frequency domain. The usual conditions apply, namely that one signal shifted in the frequency domain must not overlap its neighbours to any noticeable degree; and also the resulting combined signal must be matched to the characteristics of the line along which it is to be transmitted.

Time Division Multiplexing (t.d.m.). In pulse amplitude modulation the width of the sampled pulses is made sufficiently narrow so that other p.a.m. signals can be fitted into the time gaps. The mechanism illustrated in Fig. 8.1 shows four analog signals t.d.m.'d together, to give an analog signal output. A similar but more complicated mechanism is needed for digital signals.

If there are N channels to be multiplexed then the number of pulses per second is N times that in any one p.a.m. signal. This resulting signal must be matched to the characteristics of the line along which it is to pass.

Space Division Multiplexing is a term used in the telecommunications industry to mean the physical packing of more than one wire into a cable, each wire being used for one channel.

Pulse Code Modulation Multiplexing is really a form of time division multiplexing. If each character is a binary number then provided the message coding is performed sufficiently quickly and the binary digits as pulses are sufficiently narrow, characters from different signals can be combined together into one signal (see ref. 8.1).

In all cases the receiver must have complete knowledge of the structure, in both the time and frequency domains, of the incoming signal in order that correct selection of the multiplexed signals can be made.

8.2 Multiplexers

Multiplexing is defined as the division of a single transmission channel into two or more channels. The equipment which carries out this task is called a multiplexer. There are a number of types of multiplexers which may be used in a computer system, apart from the data transmission area. Multiplexing techniques are used, for example, to attach a number of slow peripherals to the system, as shown in Fig. 8.2. In this case the multiplexer channel (or input/output bus) con-

DATA TRANSMISSION

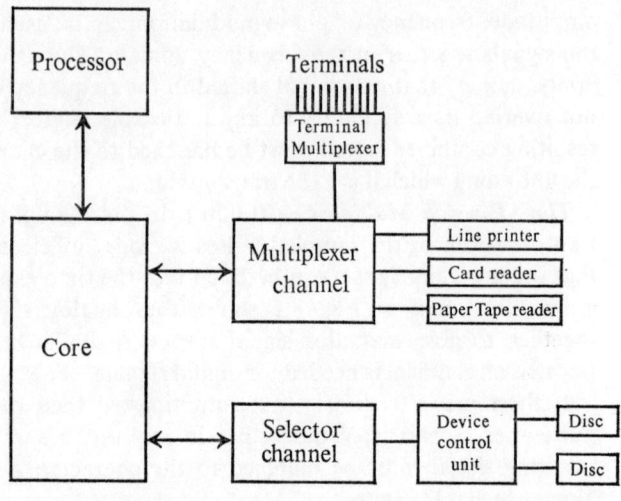

Fig. 8.2

trols the transference of data to and from the peripherals attached to it, all of which may be operating simultaneously.

Multiplexers used in data transmission control the lines used for the transmission and reception of data between a computer system and a number of terminals. The typical, simplified action of a multiplexer is shown in Fig. 8.3. It is clear that the time t must be such that even if every line under control of the multiplexer were transferring data none would be lost. The number of lines a multiplexer is able to handle is limited such that the sum of the transfer rates of the devices attached to the transmission lines does not exceed the total throughput of the multiplexer. (It may also be limited in practice by the number of access ports provided, or even by some software limitation in the computer system.) Thus if t is the time in seconds taken by the multiplexer to select a line and then to transfer one unit of data (a character, say), and that the minimum interval between the arrival of two successive units of data at the multiplexer for any given line is n seconds, then the capacity of the multiplexer is not greater than n/t lines. If the transmission speeds vary across the lines, or if it is known that all lines cannot or will not be active together, then these factors may be taken into account in determining the capacity of the multiplexer.

MULTIPLEXING AND MULTIPLEXERS

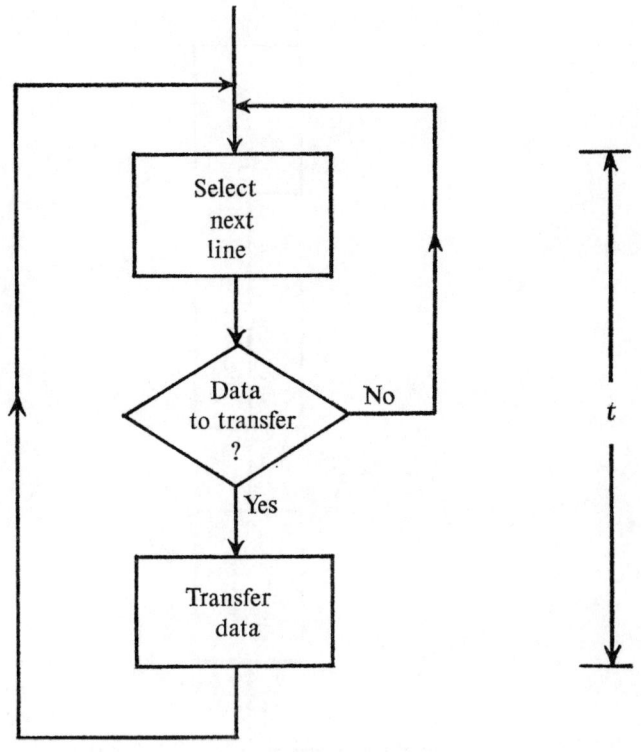

Fig. 8.3

Multiplexers may be conveniently classified into two types—those which are implemented solely as hardware, and those which make use of a computer to carry out the multiplexing and employ a combination of hardware and software.

Hardware multiplexers exhibit the following characteristics:

(i) They are built in such a way that they are only able to accomplish the limited tasks for which they were designed.

(ii) There may be a duplication of certain hardware for each line.

(iii) The line characteristics may be fixed at construction time; e.g. data rates, data format.

Multiplexers using a processor exhibit the following characteristics:

71

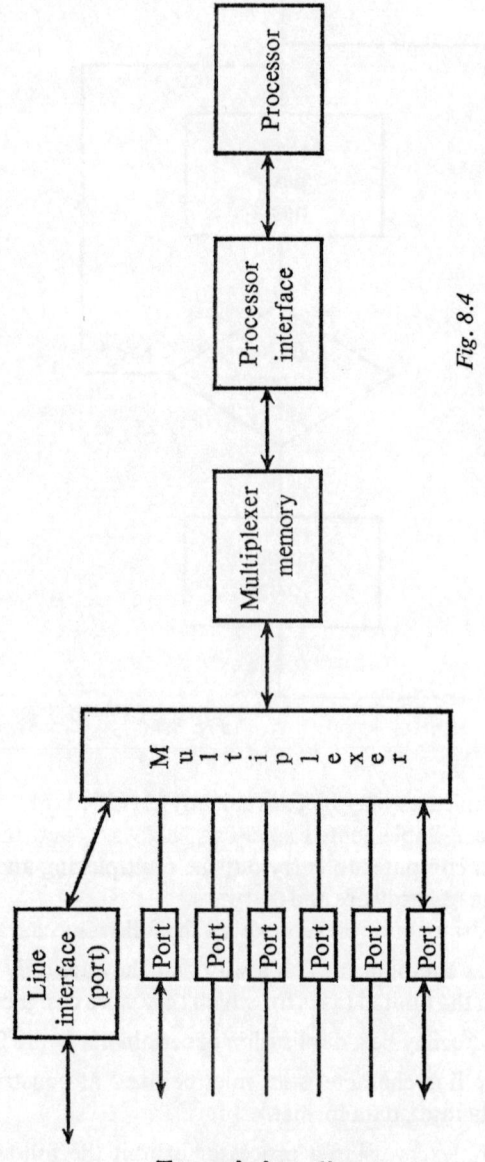

Fig. 8.4

MULTIPLEXING AND MULTIPLEXERS

(i) They have an instruction set which allows a more flexible approach to be taken to the overall communication problem.

(ii) The use of software for line scanning may save costs for a large number of lines; i.e. if the reduction in cost per line is sufficient to balance the extra cost of a processor over a simple hardware multiplexer.

Thus communication processors offer more flexibility, and may in certain circumstances offer a cost advantage.

In many cases it is not possible to isolate the equipment which carries out the pure multiplexing function. For example, the problems of interfacing each line, character synchronisation, detection of special characters or code conversion are often carried out in a single item of equipment which is given the general name of multiplexer.

In its simplest form we may illustrate the position of the multiplexer in a system as in Fig. 8.4. In this diagram the multiplexer is multiplexing the path between the memory and the transmission lines. The processor interface may be carrying out code conversion on the incoming and outgoing data and may be detecting and responding to special characters. However, it would be usual for a manufacturer to refer to everything to the left of the processor in this diagram as the communications multiplexer.

The way in which multiplexers are attached to a given system depends upon a number of factors such as cost, the application of the system and transmission speeds. Figure 8.5 illustrates three approaches to interfacing a communications multiplexer to a system. Figure 8.5(a) shows a number of low-speed lines multiplexed on to

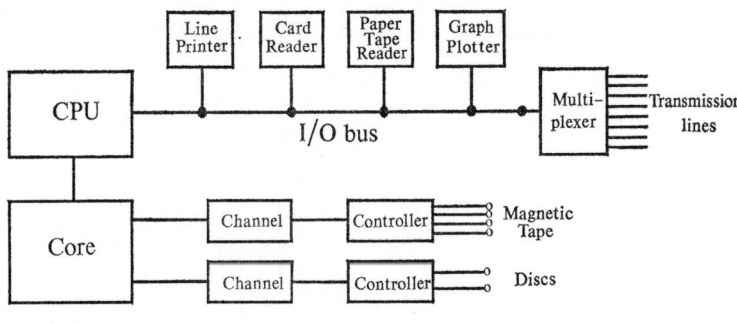

Fig. 8.5(a)

73

DATA TRANSMISSION

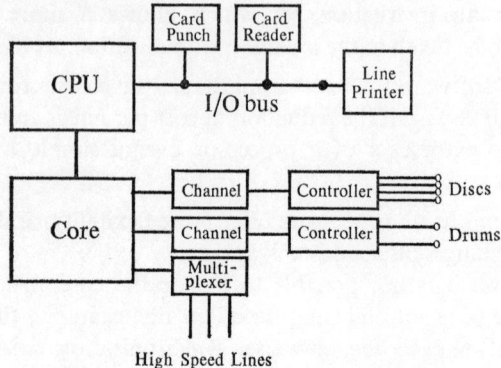

Fig. 8.5(b)

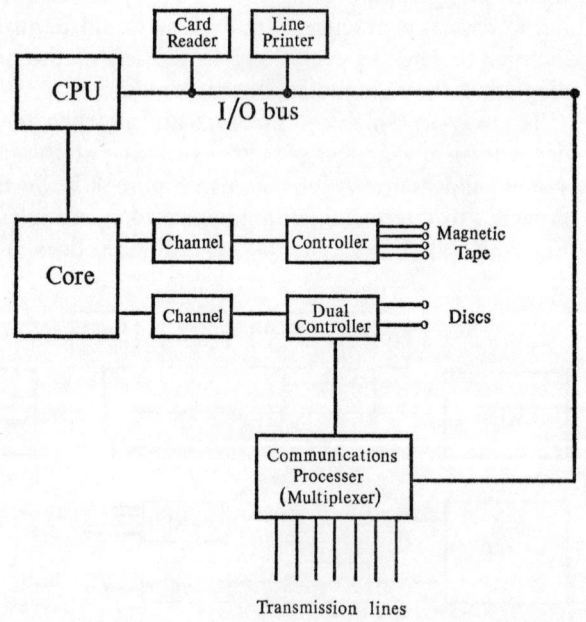

Fig. 8.5(c)

the input/output bus of a system; the devices might typically be Teletypes or similar devices. Figure 8.5 (b) shows a number of high-speed lines multiplexed directly into core. The lines might be to remote batch terminals, or to other computers. Figure 8.5 (c) shows a dual controller of a disc with the communication processor using this channel to pass data to and from the disc; control information is passed via the input/output bus of the main processor. This type of configuration might be suitable for a data-collection system.

8.3 Line sharing

In order to reduce the unit cost of long-distance line rental a number of techniques have been developed for sharing the use of such a line between a number of terminals.

Consider, for example, the case of a time-sharing bureau which wishes to offer its services in areas remote from the central computer. Clearly the bureau would be at a disadvantage to local competitors if it required the subscriber to pay for the long-distance telephone calls. To overcome this the bureau might install a *concentrator* (typically a small communications processor) in the area, to which the local subscribers would be connected by low-speed lines in the usual way. The concentrator is then connected by a higher speed line (2400–4800 baud) to the central computer. The subscribers pay for their local calls and the bureau rents the high-speed, long-distance line (recovering the cost indirectly from the subscribers). This is shown in Fig. 8.6.

The concentrator acts as a message collector for the users in the area. Messages are built-up in the concentrator's storage for each user, and on meeting a delimiter (typically carriage-return and/or line-feed) the complete message is then transmitted over the high-speed line. Together with the message is sent terminal identification. Thus the concentrator is acting as a message-switching centre and is making use of store-and-forward techniques.

The transmission over the high-speed line may be either synchronous or asynchronous. Since on a time-sharing system there is likely to be users local to the central computer, using asynchronous transmission techniques, it is often convenient to use the local com-

DATA TRANSMISSION

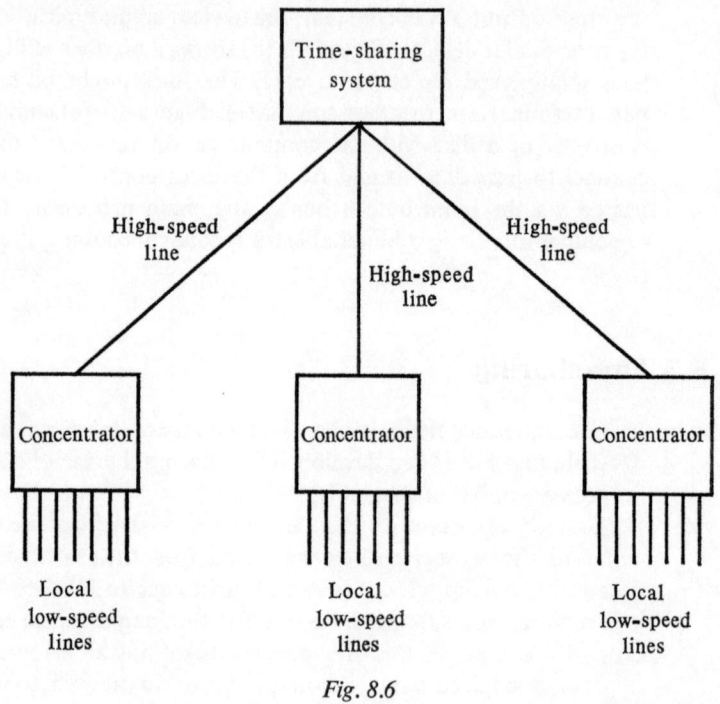

Fig. 8.6

munication subsystem to deal with the high-speed lines in an asynchronous manner. This implies that the communication system is able to deal with at least two different transmission rates. If there are a large number of high-speed lines it may pay to deal with them separately, using synchronous techniques, perhaps with a slightly higher throughput than would be possible using asynchronous methods.

Messages from the central system are sent to the appropriate concentrator (using part of the terminal address to select the correct high-speed line). The concentrator then uses the rest of the device address to route the message to the correct destination along the local low-speed line.

The concentrator must deal with such things as character echoing, message editing (e.g. delete previous character, delete entire line) and recognise special characters such as end of message. There are

MULTIPLEXING AND MULTIPLEXERS

obvious similarities with the type of communication subsystem which was described earlier in this chapter.

Another technique used to share a high-speed line is where a number of devices (terminals and/or concentrators) are attached to the same line. Such a system is shown in Fig. 8.7, and is known as a *multidrop network*.

With a multidrop network all messages are sent and received under control of the central computer (or its communication subsystem) one at a time. All devices on a multidrop line have an address which they must be capable of recognising and responding to. Messages from the central computer are preceded by a *select message*, which is the address of the selected receiving device. All devices except the selected one ignore the select message. The selected device may acknowledge (if synchronous transmission with handshaking is being used) or it may simply accept the data which follows the select message. On some multidrop systems it is possible to send data to more than one device by preceding the data by a number of device addresses or by having a broadcast address which may route the data to all the devices on the line, or if a concentrator is on the line it may route the data to all the terminals on that concentrator.

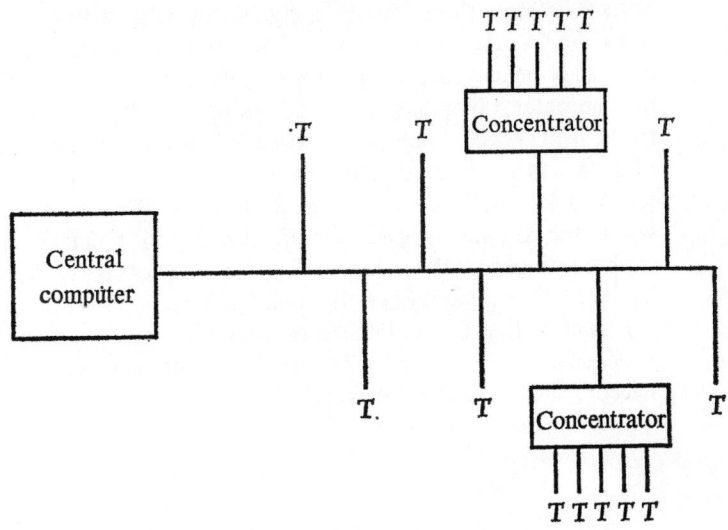

Fig. 8.7

DATA TRANSMISSION

In the same way that the transmission of messages from the central computer to the device is under central computer control so the transmission of data from the devices is also controlled by the central computer. The central computer sends out a *polling message* with a particular device address. This message is ignored by all but the addressed device. The polling message inquires of the device: 'Have you any data ready to send?' The reply is either 'no', or the data to be transmitted. If the reply is no, or after the reception of the data, the next poll to another device is sent. (The polling of devices is, of course, the basis of all multiplexing; the multidrop system is simply software multiplexing.) The organisation of the polling is centred around a *poll table* in the central computer which indicates the sequence of the devices to be polled. By placing a given device in the polling table at a number of points one can arrange to handle devices of varying speeds and also reflect any relative priorities among the devices. For example, if there is a concentrator on a line together with individual terminals, then to reflect the fact that the concentrator would have a much heavier traffic of data than an individual terminal one would put the concentrator in the polling table a number of times thus asking it more frequently if it had data to transmit.

In some cases, where the distance between the computer and the devices is large, there may be a significant delay between the computer sending the poll and its receiving a 'no'. In such cases the network may be designed so that rather than a device sending a 'no' to the computer it increments the address of the poll and sends the poll to the next device. Each device will need special circuitry to accomplish this. If no device has any data to transmit then the last device sends back a 'no' and the process starts again. If at any stage data is sent to the computer the device is identified in the poll table and polling is continued by the polling software to the next device in the polling list. This technique is known as *hub polling*.

It is clear that a communication processor could be employed to good effect to handle both the line interfaces and to organise the selection and polling of the lines.

9 Asynchronous line interfaces

9.1 Hardware interface

In attaching to a computer any device which uses serial transmission techniques, it is necessary to convert the serial bit stream into a parallel bit stream in a register for transmission along the computer's input/output bus. This is shown in Fig. 9.1. The equipment which performs this task is called the line interface. In the case of asynchronous serial transmission, the interface is also responsible for synchronising each character. On reception, this involves the detection of the start of the character and, subsequently, the assembly of each bit of the character and the detection of the end of the character.

The start of each character is defined by the leading edge of the start-element. Using this as reference, and knowing the bit rate for the transmitting device, the interface is able to sample the state of the line for each bit of the character. The end of the character may be determined either by counting the number of bits received or by detecting overflow of the buffer (register) in which the character is being assembled.

To detect the beginning of a character, the interface has a bistable (or flip-flop) which records the current status of the line. The line is said to be idle until the start of a character is detected, at which time it becomes active until the end of the character is detected, when it reverts to the idle state. If the bistable shows that the line is idle and the line state goes from the 1-state to the 0-state, the start-bit of the next character is indicated. If the bistable shows that the line is active, then a change from the 1-state to the 0-state merely indicates that the previous data bit was a 1 and the next a 0.

In order to sample the line at times which correspond to each bit of the character, it is necessary to use a clock whose rate is a multiple of the bit rate. It is theoretically possible to use a clock which has a frequency equal to the bit rate, but since most serial asynchronous

DATA TRANSMISSION

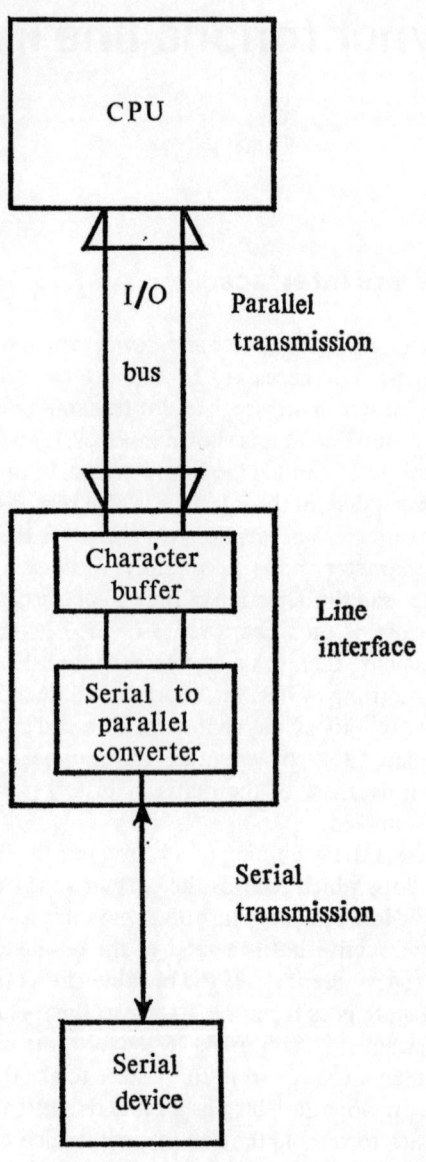

Fig. 9.1

ASYNCHRONOUS LINE INTERFACES

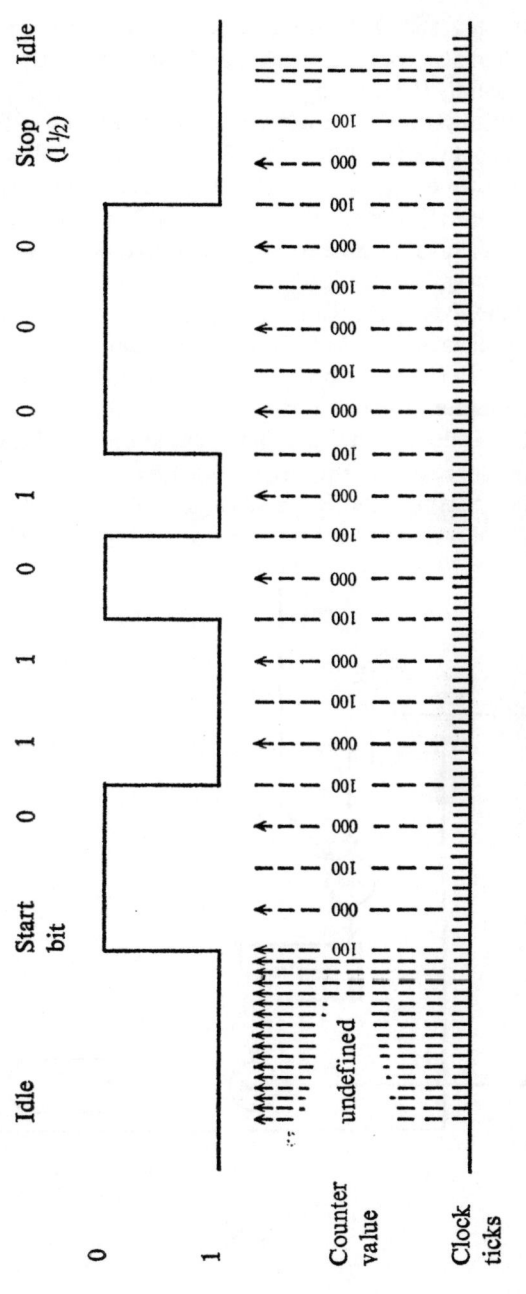

Fig. 9.2

Note: The arrows indicate the points where the line state is sampled

DATA TRANSMISSION

devices are slow such a low-speed clock is difficult to implement and has an inherent variation which may cause the sampling time to drift significantly (i.e. drift by more than the duration of one bit) and is therefore unsuitable. To overcome this limitation a faster clock may be used to drive a counter. Sampling takes place when the counter register overflows. This has the effect of dividing the clock rate by n, where n is the maximum value which it is possible to hold in the register. Thus if a three-bit register is used overflow occurs after eight clock 'ticks'.

If there are d data bits per character, then, following the detection of the start-element, the line state must be sampled at least d times. To minimise the effect of both inaccuracies of the pulse length (due to the transmitter or line distortion) and variations in the clock itself, the interface is designed to sample the line at times which correspond to the centre of each bit. Thus with a three-bit counter incremented at eight times the bit rate, if on detection of the start-element the

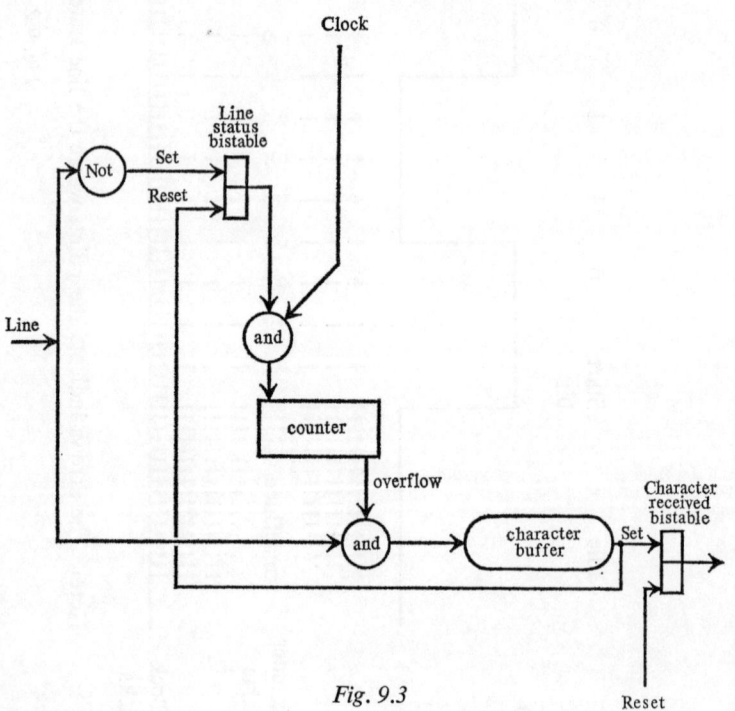

Fig. 9.3

counter register is set to the value four, then sampling will take place at the centre of the bits (to within a certain percentage). Figure 9.2 shows the state of the counter register at the various stages of receiving a character. Sampling takes place when the counter has the value zero (i.e. when overflow has occurred). Logic to achieve this is shown in Fig. 9.3 and the action of the interface at each clock tick is shown in the flowchart in Fig. 9.4.

The detection of the end of the character is achieved by inserting a 1-bit in the character buffer in such a position that when the complete character has been assembled (by shifting the register one position right for each bit of the character and inserting the line state in the most significant position), this marker bit will be shifted out of the least significant end and be detected. The detection of overflow from the buffer is used to reset the line status bistable to the idle state and to set another bistable to indicate that there is a character to pass on from this particular line.

The transmission of data from an interface to a device is carried out in a similar way. The bits of the character, including start and stop bits, are placed in a buffer and at appropriate times (obtained from a suitable clock) the line state is set to 0 or 1 depending upon the value of, say, the most significant bit in the buffer. (If the buffer shifts left it will be the most significant bit and if it shifts right it will be the least significant bit.) If the interface is to operate in half-duplex mode only, then much of the character receive logic may be used without duplication. If the interface is to work in full-duplex mode, then items such as the buffer and the three-bit counter register will be duplicated and very little other than the clock can be made common (assuming that the transmit and receive rates are the same).

In the case of the receive interface it was possible to detect the end of a character by noting overflow from the buffer. In the case of the transmit interface it is necessary to count the bits transmitted and this requires a separate register. The logic for a transmit interface is shown in Fig. 9.5 and the flowchart of the action of such an interface at each clock tick is shown in Fig. 9.6. Figure 9.7 shows a combined interface for half-duplex working.

It should be noted that in half-duplex working, if an attempt is made to transmit data in both directions at once (assuming that an interlock does not prevent this), then the effect will be corruption of the data. If data is being transmitted from a computer to a terminal

DATA TRANSMISSION

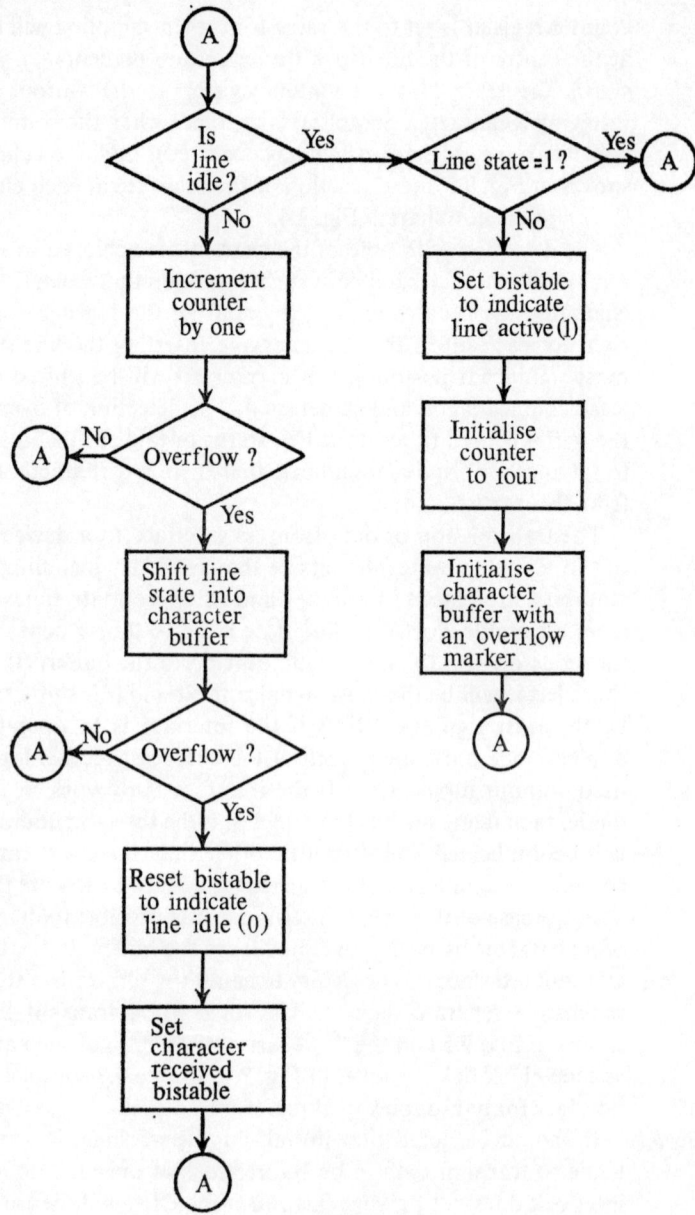

Fig. 9.4

ASYNCHRONOUS LINE INTERFACES

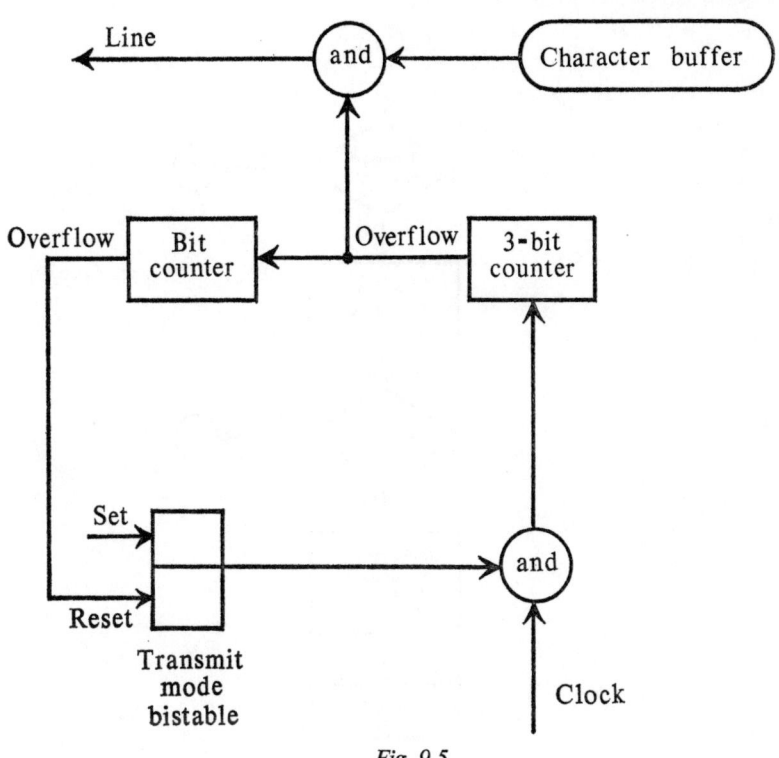

Fig. 9.5

and the terminal attempts to send data to the computer at the same time, then the effect will be corruption of the data being sent to the terminal and, since the interface will be set to transmit mode, the data from the terminal will not be recognised by the interface. This situation presents a problem in a time-sharing system where, if a user gets into a program loop in which there is output, he will be unable to send a character to the system to break in and stop the program. To overcome this, such systems often change the interface from transmit mode to receive mode every so often (say at the end of each line of output) for a short time to allow the user to interrupt.

When designing an interface for handling a number of lines, a significant cost saving may be achieved by the elimination of elements common to each interface which can be shared. If all the lines operate at the same speed (or, to be more precise, if all the devices

DATA TRANSMISSION

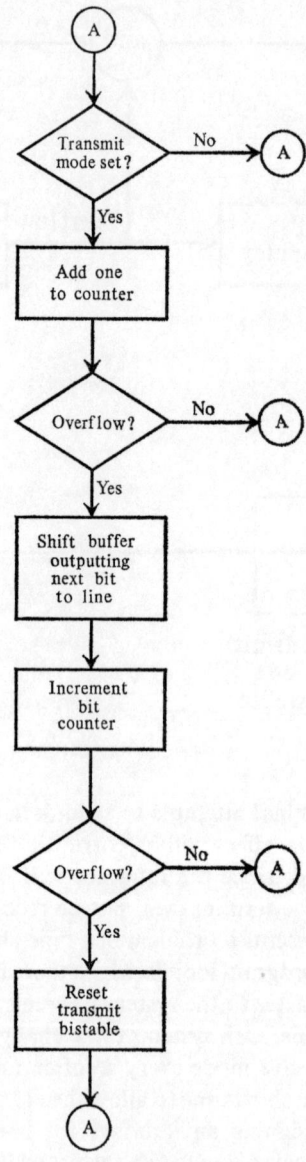

Fig. 9.6

ASYNCHRONOUS LINE INTERFACES

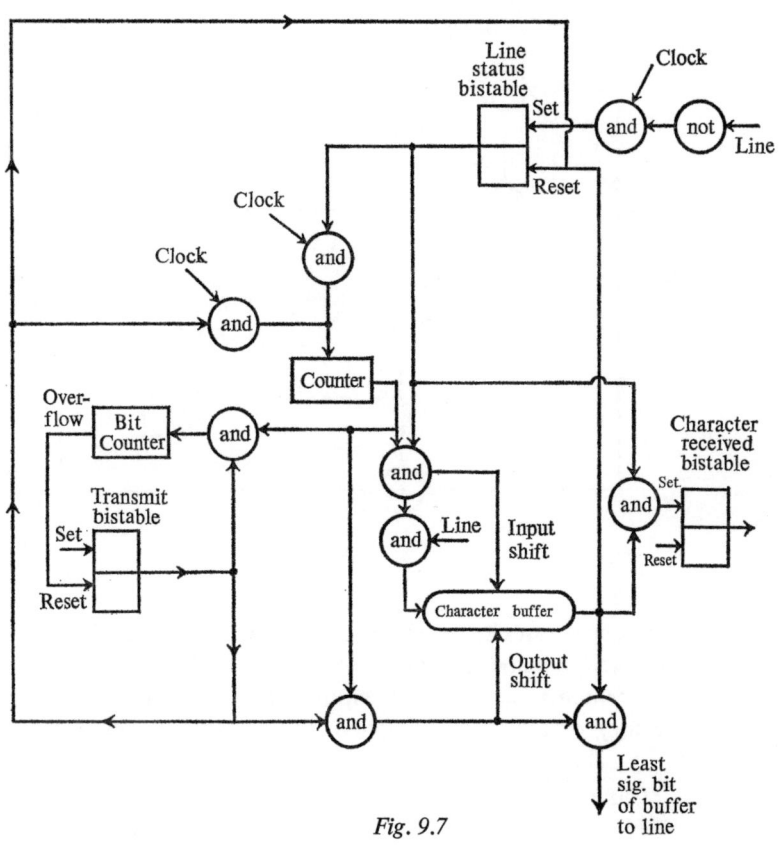

Fig. 9.7

attached to the lines operate at the same speed), then the same clock source may be used by all the lines. Above a certain number of lines a cost saving may be achieved by replacing the individual character buffers by a common memory into which characters from all lines are assembled and from which all characters are transmitted. Figure 8.4 shows such a system of line interfaces and multiplexer with a common memory. The logic of such an interface is shown in Fig. 9.8. Note that start bit detection, synchronisation and end-of-character detection remain as hardware unique to each line. As was noted earlier, the multiplexing may be carried out by a small computer. In this case the common memory is that of the small computer which

DATA TRANSMISSION

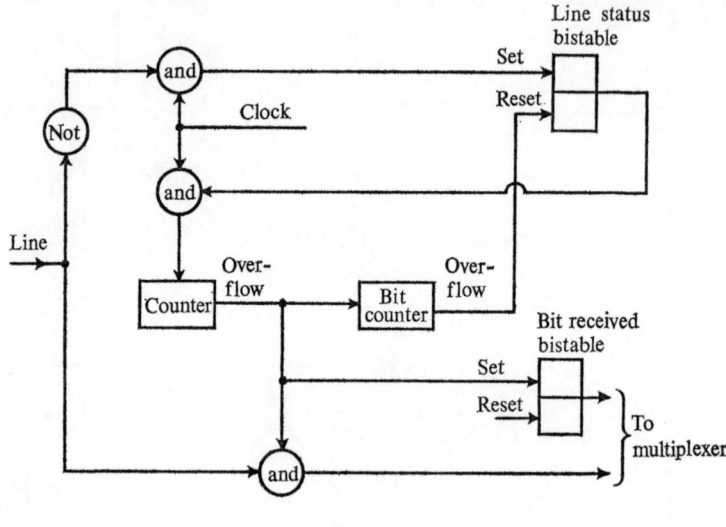

Fig. 9.8

also carries out character assembly from the bits received from the line interfaces and handles communication with the large processor. The contents of the single-bit buffers can be transferred to the memory of the small computer either under program control or using a channel direct to memory.

A significant cost saving may be possible if the small computer is used to detect the start bit, synchronise the bits and detect the end of each character. In this case the line interface is considerably simpler and herein lies the saving. However, it should be noted that the computer being used should be sufficiently fast to enable a large number of lines to be handled to make sufficient savings. It is also convenient if the instruction set is designed for communication work.

9.2 DEC PDP8

Let us consider, as a specific example, the DEC PDP8 used as a communication processor (see reference 9.1). The system operates in a transfer-in or transfer-out mode. The three instructions relevant to this are:

ASYNCHRONOUS LINE INTERFACES

(a) TTO This acts on the accumulator AC and the selected output side of the line. It causes the contents of the AC register to shift one position to the right and transfer the rightmost bit to the output line.
(b) TTINCR This increments the line select register (LSR) in the multiplexer.
(c) TTI This is used to transfer the state of the line to the computer. This instruction is complex and involves character assembly, line selection and uses three memory locations.

Following each TTI instruction are three words thus:

Location N   TTI
$N+1$ LSW   Line status word
$N+2$ CAW   Character assembly word
$N+3$ JMS   Jump to subroutine
$N+4$ TTI   Next TTI instruction

The line status word has the following format.

bit 0 is the active bit. When bit 0 is a 1, it indicates that the line was active during the last service cycle.
bit 1 is not used.
bits 2–8 give the line number of the line being sampled (maximum of 128 lines—7 bits).
bits 9–11 contain the line sample count (LSC) which is used to count the number of clock ticks up to five (in the last section a count of eight was used as an example) before entering a major cycle.

The line sampling procedure is illustrated in Fig. 9.9. When the start bit is detected (the line state = 0) it becomes one of the three necessary inputs to the active gate. A second input, the line selected, is only present when that particular line has been selected. The third input (active = 0) is a necessary condition. If active = 1 (indicating that the service routine has already started), the active gate will not turn on again until the active bit has been cleared to zero. The output of the active gate is used to set active = 1 (bit zero of LSW). The zero through four count is provided by the line sample counter (bits 9–11 of LSW). Once active = 1 and the line is selected again, the line sample counter (starting at 0) is incremented by one. When it reaches

DATA TRANSMISSION

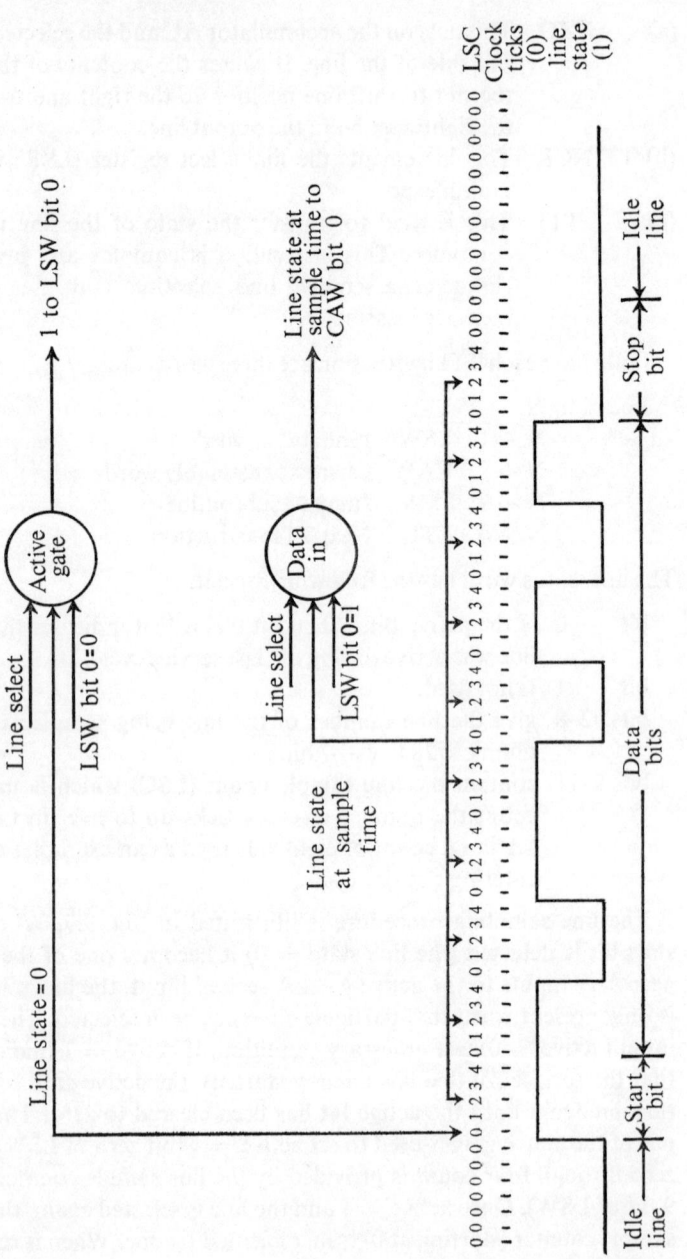

Fig. 9.9 PDPS Line sampling procedure

two (i.e. when bit 10 is set to one), the line state becomes an input to the 'data in' gate along with line select and active (= 1). When the counter reaches 4 (i.e. when bit 9 is set to one), the counter is cleared and begins counting again when the line is selected and if active is one. When the complete character has been assembled (see later) the JMS (three locations on from the TTI instruction) is executed and this, by software, clears the active bit and the line sample counter.

Line sampling is usually carried out on a block of lines using the appropriate number of 4-word TTI blocks. Because the lines are sampled at five times the bit rate (the LSC runs from 0 to 4), it is not necessary to service all characters which have gone to completion during a single pass through a block (a time-consuming action), nor is it necessary to service all possible characters during one sample time. This reduces the peak load. A load-distribution counter in the multiplexer makes this type of servicing possible.

Before a block of TTI instructions are executed, the load distribution counter is loaded with the two's complement of the number of lines to be serviced this pass (normally less than the total number). Each time a character is completed (which causes a jump to the subroutine to process it), the counter is incremented by one by the subroutine. When the counter reaches zero, the hardware inhibits the execution of further JMS instructions by taking the next instruction from TTI + 4 (i.e. the next TTI instruction in the block unless the end of the block has been reached) rather than from TTI + 3 which would be the normal event when the character has been assembled.

If a JMS would have been executed and the load distribution is zero, a line hold flag is set unique to this line. Servicing of this character will be completed during another service cycle. The line hold flag will be cleared by the TTI on some other pass whenever the load distribution counter is not zero.

The line hold flag is a flip-flop unique to each line and is used to record when a completed character has been detected, but a JMS has not been executed (because the load distribution counter was zero). The effect of the line hold flag and the load distribution counter is shown in Fig. 9.10. Whenever a new TTI instruction is to be executed, the logic looks at the line hold flag to determine if character servicing on the last pass was not completed. If the line hold flag is zero, a JMS was not inhibited on a previous cycle and the logic checks the character complete flag (bit 11 of the CAW—see

DATA TRANSMISSION

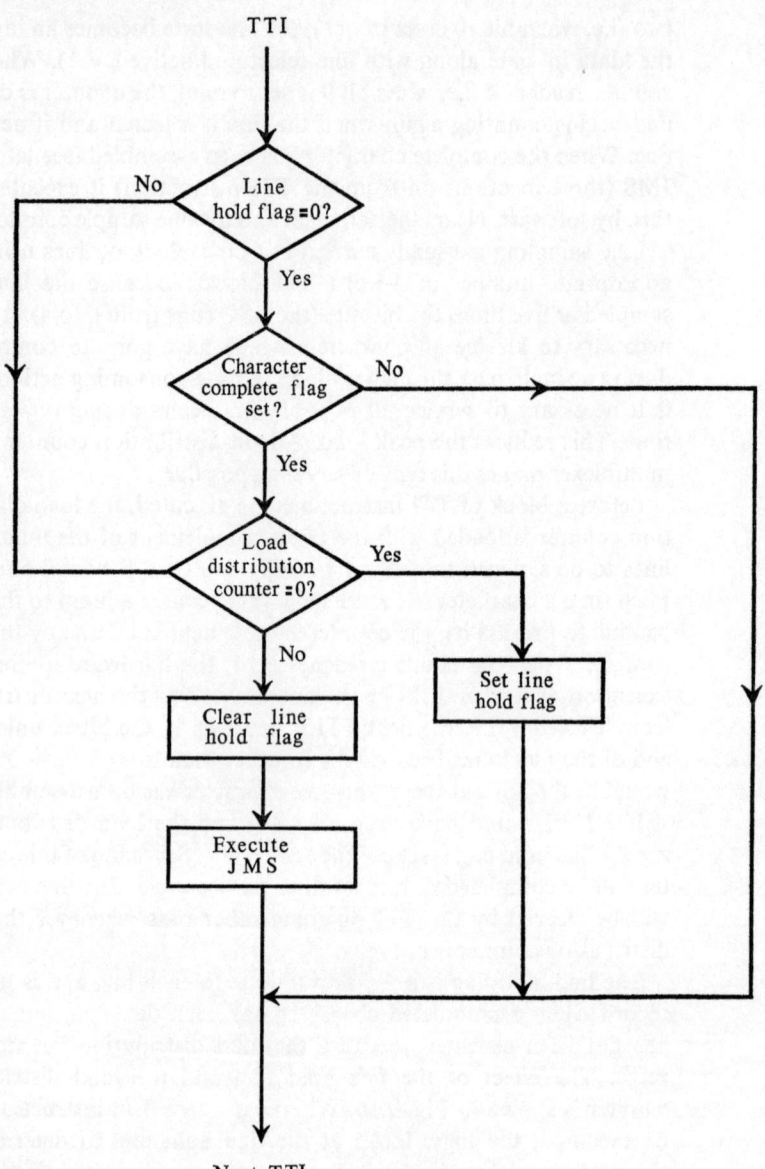

Fig. 9.10

ASYNCHRONOUS LINE INTERFACES

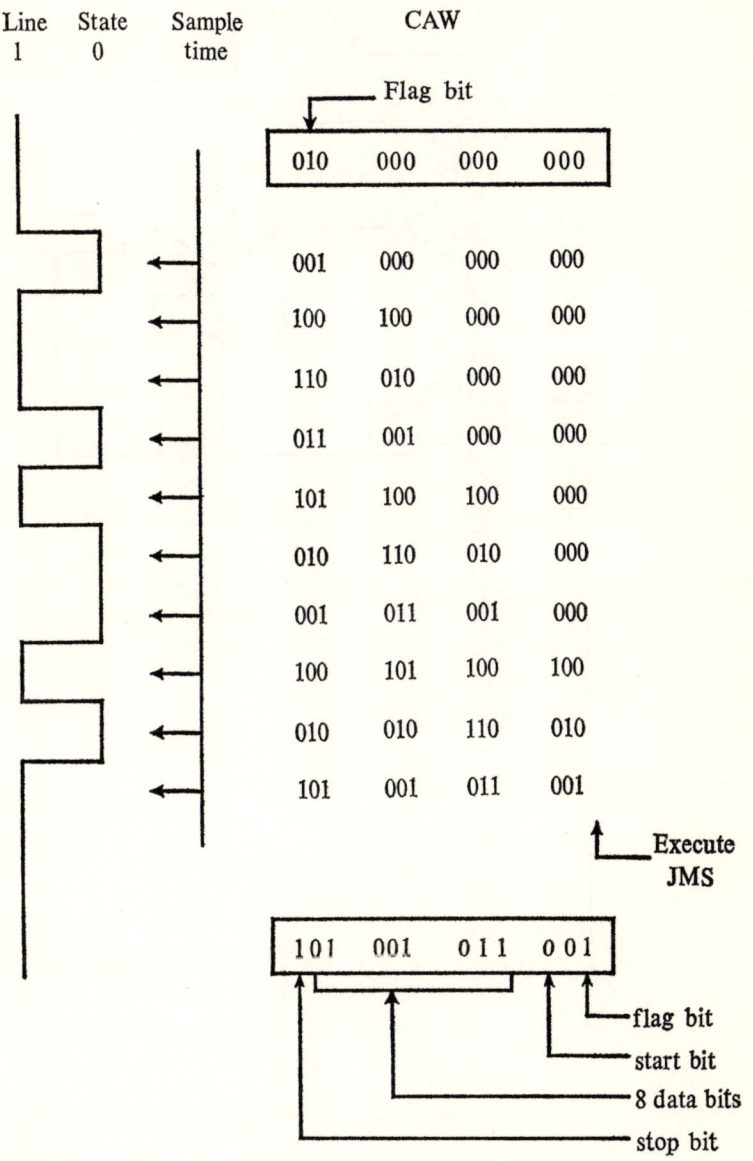

Fig. 9.11

DATA TRANSMISSION

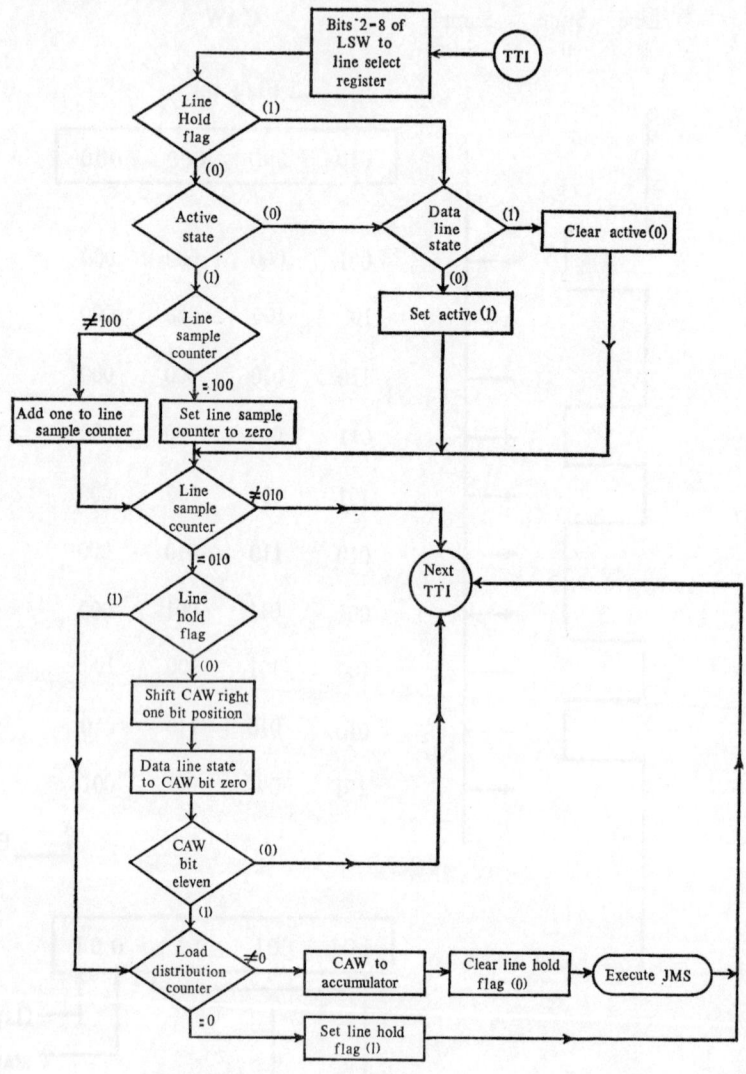

Fig. 9.12

ASYNCHRONOUS LINE INTERFACES

later). If the line hold flag is one, a requested JMS was not executed during the last service cycle and the logic looks at the load distribution counter to see if there is time on this cycle to service the character.

If the character complete flag is not set then the next activity will be the next TTI instruction. If the flag is set, then the load distribution counter is examined to see if there is time to service the character. If the load distribution counter is zero there is insufficient time to service the character and the line-hold flag is set and the next TTI is executed. If the load-distribution counter is non-zero then the line-hold flag is cleared and the JMS instruction executed.

Character assembly begins when a selected line has been found active (i.e. the start-bit has been detected). The character assembly word is illustrated in Fig. 9.11. When the line state of an inactive line becomes zero, a flag bit is placed by software into bit 1 of the CAW. On the next clock sample (on the count of 2 in the line sample count in LSW), the CAW is shifted one place to the right and the state of the line is brought into bit zero of the CAW; the data bits are assembled in the same manner. Each time a shift operation is completed, the logic looks at bit 11 to determine if the flag bit has reached this position yet. When bit 11 is set, the program attempts to execute a JMS.

The complete action of the TTI instruction (together with its effects upon the line status word and the character assembly word) is illustrated in Fig. 9.12.

The subroutine to which the JMS branches stores the CAW, stores the line number, clears the active bit and the line sample counter in the LSW, resets the CAW to contain only the flag bit in bit one, increments the load-distribution counter and administers the buffers into which the character is being placed.

It is clear from the above how powerful the TTI instruction on the PDP8 is. This is an excellent example of a computer with an instruction set designed for communication work. The PDP8 is not the only such machine but was chosen as an example since it is in such widespread use.

9.3 Main processor software

In this section the software required in the main processor to deal with the communication subsystem is examined. The software in-

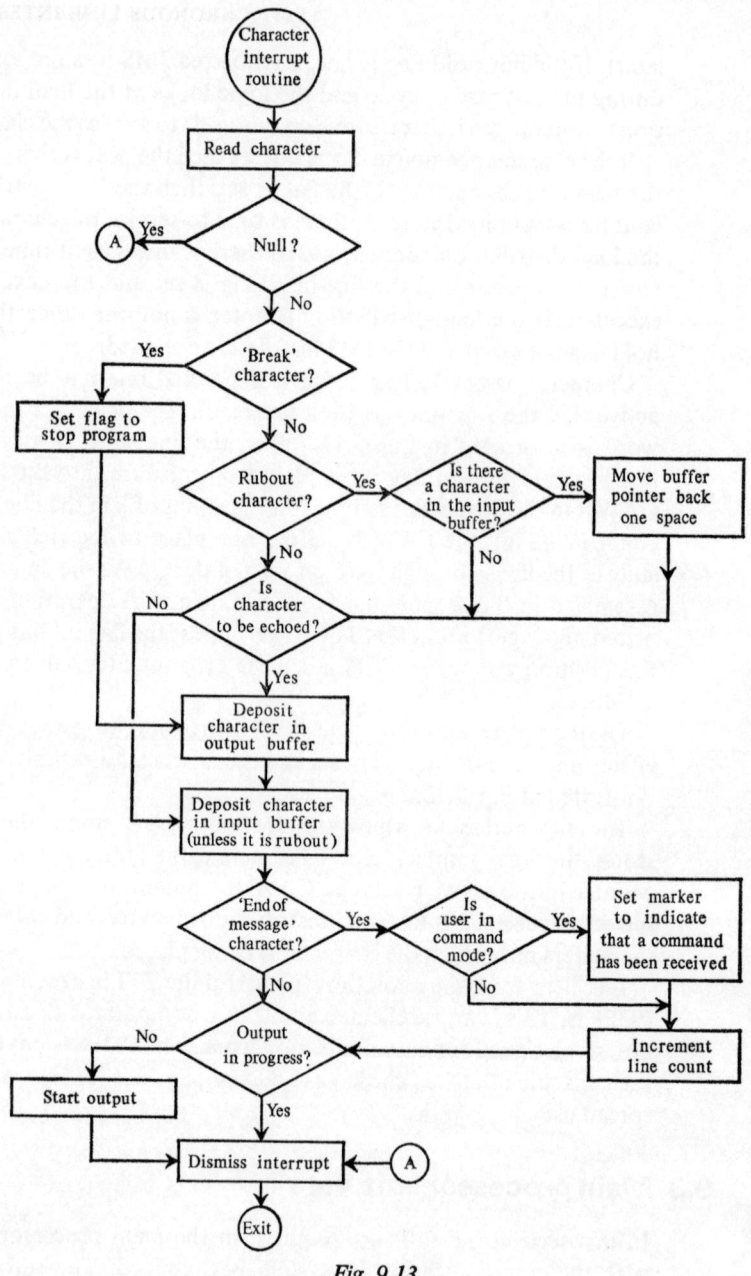

Fig. 9.13

ASYNCHRONOUS LINE INTERFACES

volved is almost always integrated into the operating system software and also reflects the application rather than the type of transmission being used (which is obviously a function of the application). Thus although both synchronous (see Chapter 10) and asynchronous interfaces pass on a single character at a time, the way in which the character is handled beyond the interface differs considerably according to the application.

Consider for example a time-sharing system using a communication processor to handle the terminals. The communication processor may pass either a single character at a time to the main processor or may pass a group of characters—typically a line of input. In the former case, the main processor will be involved in building the characters received into meaningful units such as lines of text. If the communication processor is able (fast enough and has a large enough memory) to handle the building of lines of text then this will obviously release main processor time for more useful work. It should be noted, however, that some time-sharing systems react on a character-by-character basis rather than a line-by-line basis and in this case the communication subsystem must pass single characters.

The routine in the main processor to respond to an interrupt from the communication subsystem to pass a single character is illustrated in Fig. 9.13. (The case of passing a line of text is obviously a subset of this since much of the work will have been carried out by the communication processor.) Assume that the system is operating in full-duplex mode and that since some characters, such as passwords, will not be echoed, the main processor will handle character echoing.

Obviously most of the work should be done if at all possible by the communication subsystem but this represents the 'worst case' of main processor loading. The flowchart does not deal with the overall problem of buffer management and things such as 'type-ahead' where the user is entering information into the input buffer before the system is ready to deal with it.

Notice the application dependent features such as the 'break' character to interrupt a running program and return to command mode, the rubout to delete the previous character, the treatment of the end of message when in command mode and the test for non-echoed characters.

10 Synchronous line interfaces

10.1 Hardware interface

The use of synchronous as opposed to asynchronous techniques permits a higher rate of data transmission over a given channel. To utilise this higher transfer rate, data is transmitted in conveniently sized blocks and this necessitates the use of buffers to hold the data blocks prior to transmission. This also allows error recovery procedures to be designed and implemented.

Synchronous modems supply all timing necessary to send or receive each bit as it is made available to or from the modem. The line interface for a synchronous modem is somewhat simpler than that for an asynchronous line; for example, synchronisation is less complicated. The main area of difficulty is to incorporate the capability of communicating in the message formats used in synchronous transmission. It is possible to design a hardware interface which will check and filter the control characters, carry out block parity checks and respond appropriately to error situations. Such an interface is almost bound to be format sensitive and as such will have a limited market. To build-in flexibility to a hardware interface may cause it to be too expensive. An alternative for a number of lines is to make use of a communication processor (mini computer) which will handle by software the message formats and error conditions.

There are two types of information transmitted in a synchronous transmission system—data messages and control characters. The control characters are used to enable the system to discover the status of elements within the system, separate data messages and recover from an error situation.

The various control characters used for message control, message heading and error detection and correction were described in Chapter 2 (see Fig. 2.2). Of these characters, five are what is known as 'turn-around' characters. These characters, ETX, ETB, ACK, NAK

SYNCHRONOUS LINE INTERFACES

and ENQ, cause the elements of the system (transmitter or receiver) to change their role either after transmitting or receiving such a character. Thus when the transmitter sends ACK, it will not transmit again until it has received a reply from the receiver. Similarly, the receiver when it receives the ACK will then change its role and transmit a reply. The characters which the receiver transmits will contain a turn-around character so that after sending the reply it will await information from the transmitter. This technique is known as 'hand-shaking'.

In order to establish synchronisation, a synchronising sequence (typically three SYN characters) is placed at the beginning of every transmission.

Before the first data message is transmitted it is usual to establish the status of the receiving station by sending an ENQ character. The reply to an ENQ is usually a character or group of characters unique to the receiving station and serves to identify the station. The identification characters may be followed by a NAK character. Only when the NAK has been received and the identification characters checked will the transmitter send the first message. The situation where no reply is received will be discussed later.

The transmission block size is governed by the buffer size of either the transmitter or the receiver. All transmission blocks are terminated by parity characters which usually take the form of a longitudinal parity check. The various data message formats are as follow

(a)   synchronising sequence
        SOH
        header information
        STX
        data
        ETX
        parity character

In this case the entire message fits within one buffer

(b)   synchronising sequence
        SOH
        header information
        STX
        data
        ETB
        parity character

DATA TRANSMISSION

In this case the entire message would not fit into the buffer and the message is broken into a number of blocks terminated by ETB. except for the last, which would be terminated by ETX. In some situations the header information may be unnecessary, in which case the SOH character and the header information would not be transmitted.

Following transmission of a message, the message is retained in the transmitter's buffer until such time as an ACK character is received by the transmitter indicating the successful reception of the message.

There are basically two types of error situations that can arise. Either the receiver detects a parity error within the message or the receiver fails to reply. (Note that the receiver in both cases could be the transmitter turned around awaiting a reply from the receiver.) If a parity error is detected the receiver will reply with NAK instead of ACK, in which case the transmitter will retransmit the contents of its buffer. Only on receiving ACK will the transmitter overwrite the contents of its buffer with the next message. If there is no reply to a message then one of two control sequences are transmitted at regular intervals until either a reply is ultimately received, or the system shuts down and informs an operator of the event. If the transmitter receives no reply it sends ENQ. If the receiver detects ENQ it will respond with NAK (as it did in the opening sequence) together with station information. The NAK will cause retransmission of the message. If the receiver gets no reply to an acknowledgement (and end of transmission (EOT) has not been sent) then the receiver sends out ACK at regular intervals. If the transmitter detects the ACK it will continue with the next message as if nothing had happened. The various error situations and the action taken are indicated below.

Transmitter	Receiver	Comment
Synch. seq.		Establish
ENQ		communication
→		
	Synch. seq.	
	Station info.	Receiver
	NAK	ready
←		

SYNCHRONOUS LINE INTERFACES

Synch. seq. STX Message 1 ETB Parity character		1st data block
→		
	Synch. seq. ACK	Received correctly.
←		
Synch. seq. STX Message 2 ETB Parity char.		Refill transmitter's buffer and send message 2.
→		
		No reply in a given time.
Synch. seq. ENQ		
→		
		No reply.
Synch. seq. ENQ		
→	Synch. seq. NAK	Received ENQ.
←		
Synch. seq. STX Message 2 ETB Parity char.		Retransmit message 2.
→	Synch. seq. NAK	Parity error.
←		
Synch. seq. STX Message 2 ETB Parity char.		

DATA TRANSMISSION

	→	
	Synch. seq.	Received
	← ACK	correctly.
		No reply to
	Synch. seq.	ACK so
	ACK	retransmit it.
	←	
Synch. seq.		
STX		ACK received,
Message 3		send next
ETB		message.
Parity char.		
	→	
	Synch. seq.	Received
	ACK	correctly.
	←	
.		
.		and so on.
.		
.		
.		

In many interfaces the response required in reply to an initial ENQ from the transmitter is

    SYN
    station information
    ACK

The implication of this is that the response to an ENQ in an error state is different from the initial ENQ and this complicates the interface. However, the advantage of using ACK is that the receipt of ACK at any time triggers the refilling of the transmitter's buffer and so can be used to set the transmission going, rather than needing some other mechanism within the interface to fill the buffer initially.

As with asynchronous transmission, it is attractive to use a communication processor to deal with a number of lines. In the synchronous case, the interface is fairly simple but the software to deal with message and control characters is not so simple.

SYNCHRONOUS LINE INTERFACES

10.2 Software

Consider a very simple subset of the software requirements. The flowcharts dealing with transmission/reception of a single message format are shown in Figs. 10.1 and 10.2, where the format of the messages is assumed to be:
>SYN
>SYN
>SYN
>STX
>text
>ETX
>parity character

The flowcharts assume that the processor operates in a program interrupt mode, responding to transmit and receive interrupts from the interface. The function of the various software switches are as follows.

(a)   ACK This is set when either a message has been received and the ACK or NAK is about to be sent (it is cleared as soon as one has been sent) or a message has been transmitted but ACK has not been received.

(b)   SS This is set when the first of the three SYN characters is transmitted and cleared when all three have been transmitted. It is used in a similar way by the receive program.

(c) THOLD This is set to indicate to the main program that transmission is in progress. When exit is made from the transmit or receive routines, back to the main program, THOLD is tested by the main program.

(d)   TS If TS = 0 the message is to be transmitted.
If TS = 1 ACK should be transmitted.
If TS = 2 NAK should be transmitted.
If TS = 3 the transmission has finished and the interface must be turned around.

(e)   DCTR This is set initially to the negative number of characters in the message.

(f)   DTC This is set to 1 when the data transfer is complete.

(g)   SCTR Is used to count the three SYN characters.

(h)   RCTR This is initially set to the negative size of the buffer (i.e. the maximum number of data characters it is possible

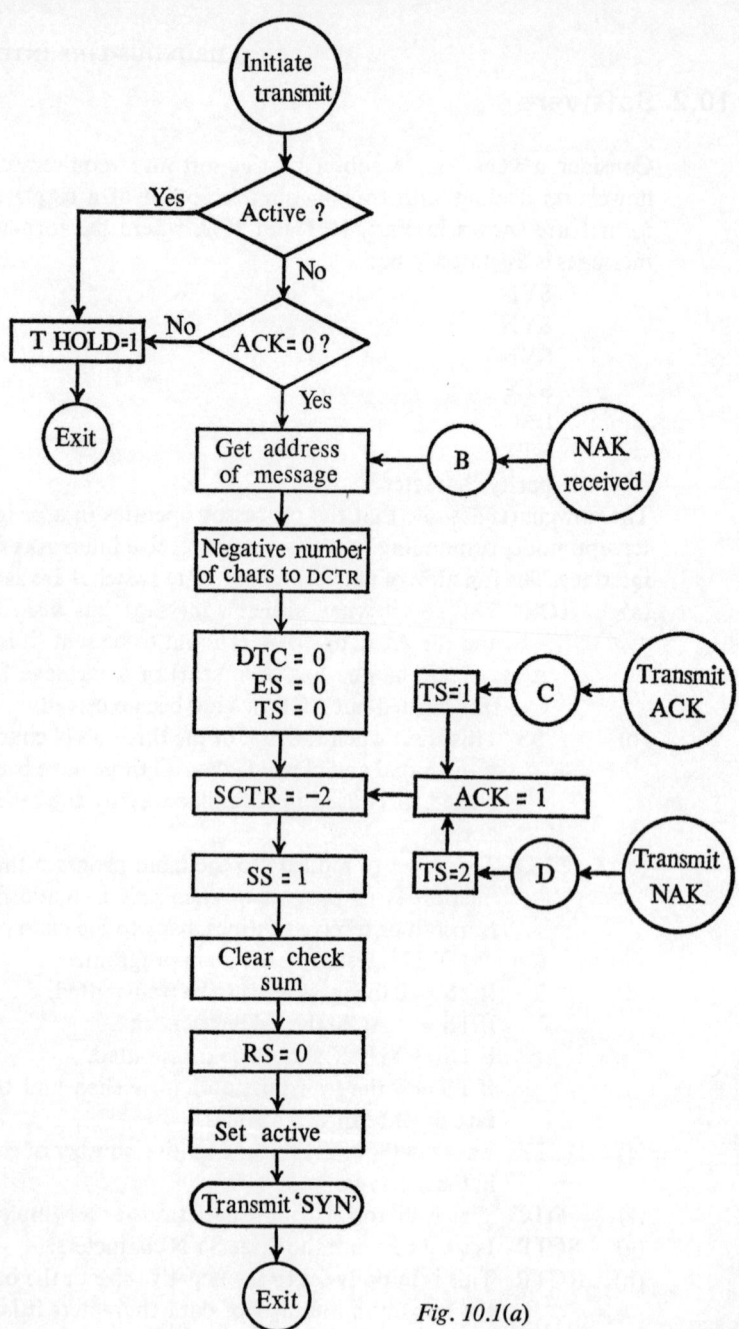

Fig. 10.1(a)

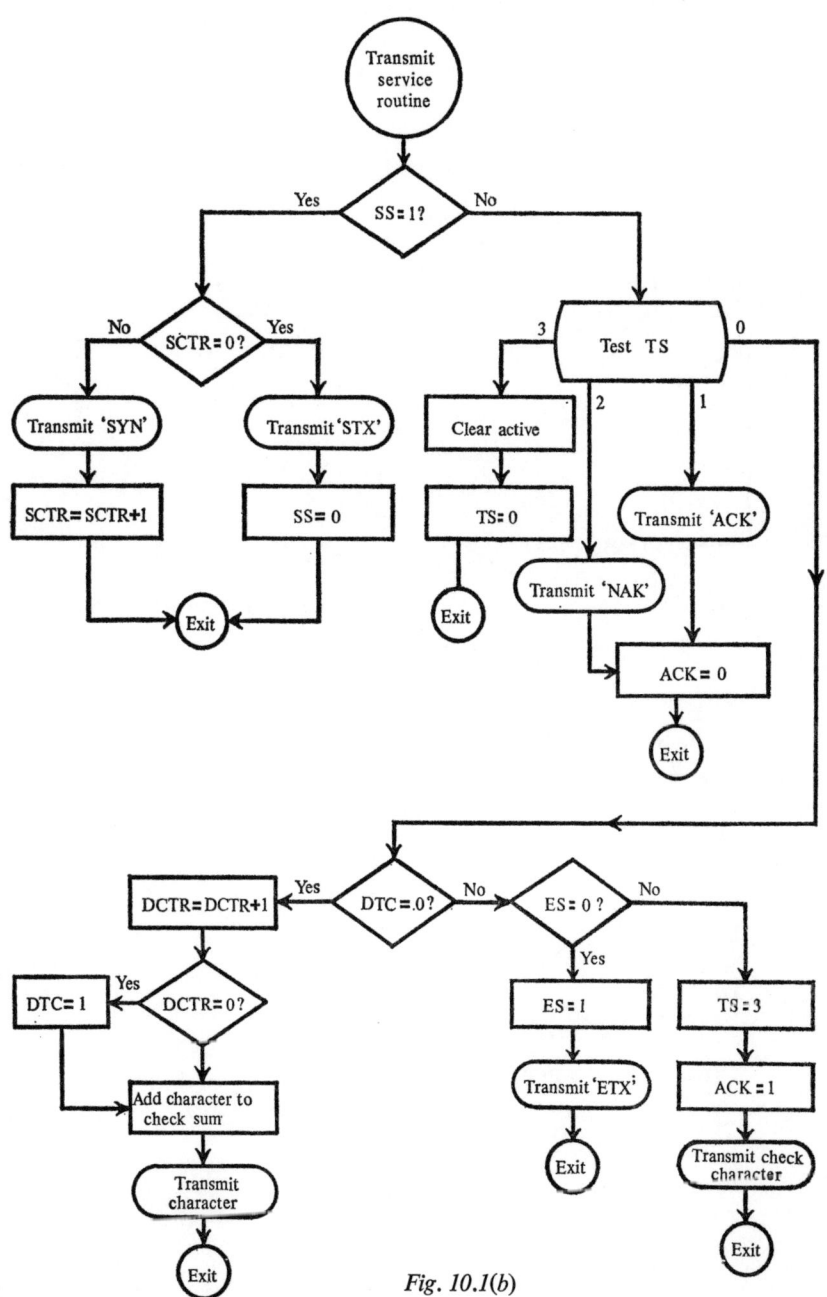

Fig. 10.1(b)

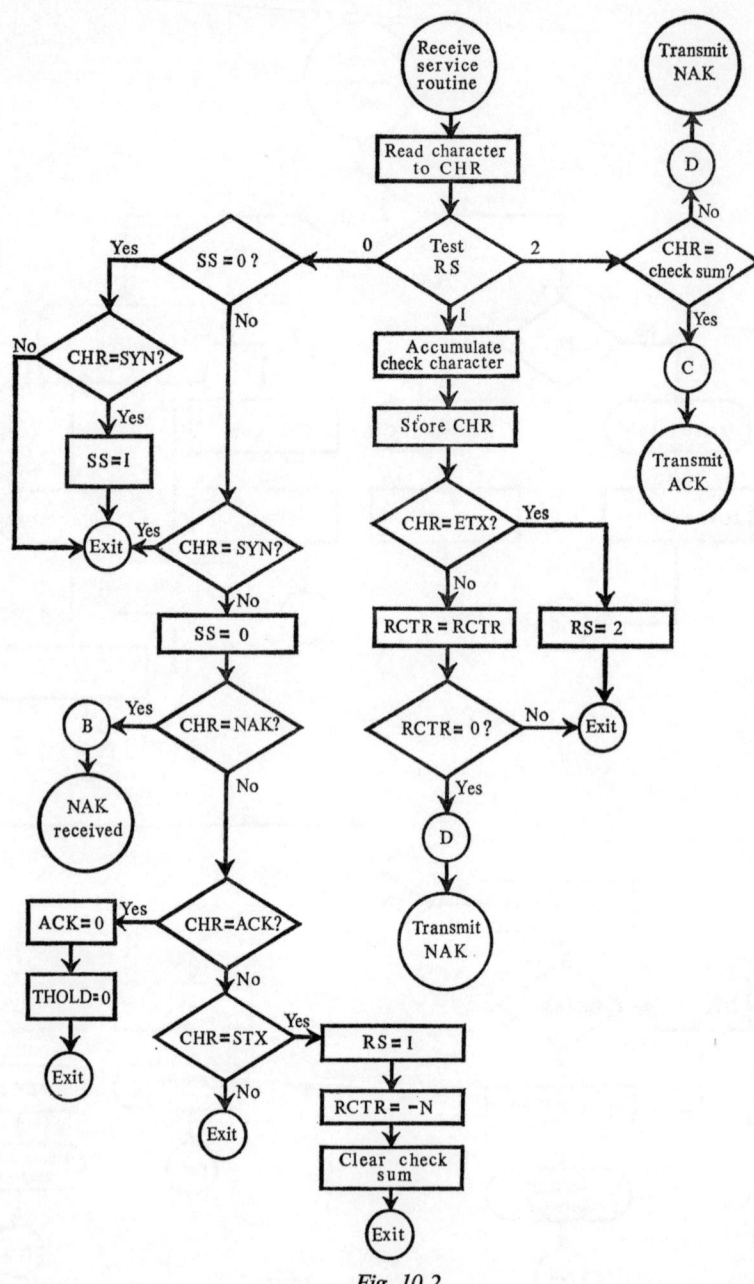

Fig. 10.2

SYNCHRONOUS LINE INTERFACES

to transmit in one block). It is incremented until either ETX is met, which indicates end of block, or until the count reaches zero, in which case ETX has failed to be recognised and NAK is sent to cause a retransmission.

(i)   RS   If RS = 0 the receiver is awaiting the synchronising sequence.
If RS = 1 the message is expected.
If RS = 2 the parity should be checked against the accumulated parity count.

(j)   ES   If ES = 0 the transmitter is about to send ETX.
If ES = 1 the transmitter is about to send the parity check character.

11 Analog and parallel interfaces

11.1 Introduction

In interfacing an analog device to a digital system there is obviously the problem of analog-to-digital conversion. If the device is also an output device there is the problem of digital-to-analog conversion.

11.2 Analog-to-digital conversion

A simple A–D converter uses a linearly increasing waveform as a reference voltage in the conversion process. The input interrupt pulse is used to initiate generation of the reference waveform and to admit clock pulses into a counter. Assuming calibration of the clock and

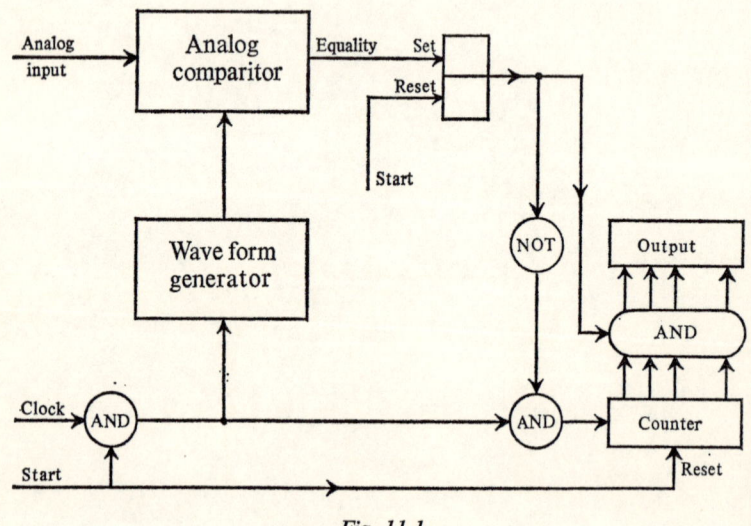

Fig. 11.1

ANALOG AND PARALLEL INTERFACES

the waveform, the value of the counter will be proportional to the reference voltage at any given time, with a maximum error of one half of one unit. The reference voltage and the analog input are sent to an analog comparitor which generates an output pulse when the two voltage are equal (or differ by a fixed small amount). The output from the comparitor is used to turn off the clock and reset the reference voltage. The counter then contains the converted value (or some fixed proportion) of the analog input. The interface is shown in Fig. 11.1.

This simple form of converter is very slow and other faster techniques are preferred. If the digital input register has n bits, then the above method requires 2^{n-1} clock pulses on average. A faster technique, described below, involves the use of a D-A converter which is described first.

11.3 Digital-to-analog conversion

A simple D-A converter uses a set of stable resistors weighted R, 2R, 4R, 8R, etc. These are switched into an adder under control of the register containing the digital signal. Figure 11.2 shows this converter.

The D-A converter may be used as the basis of an A-D converter in the following way. Starting with only the most significant bit set in the digital register, a D-A conversion is carried out to produce a

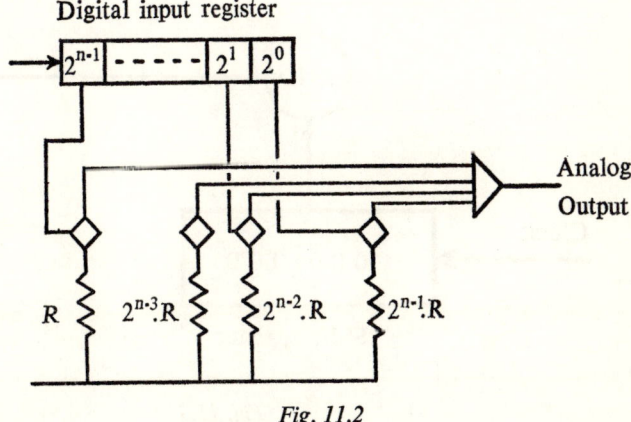

Fig. 11.2

DATA TRANSMISSION

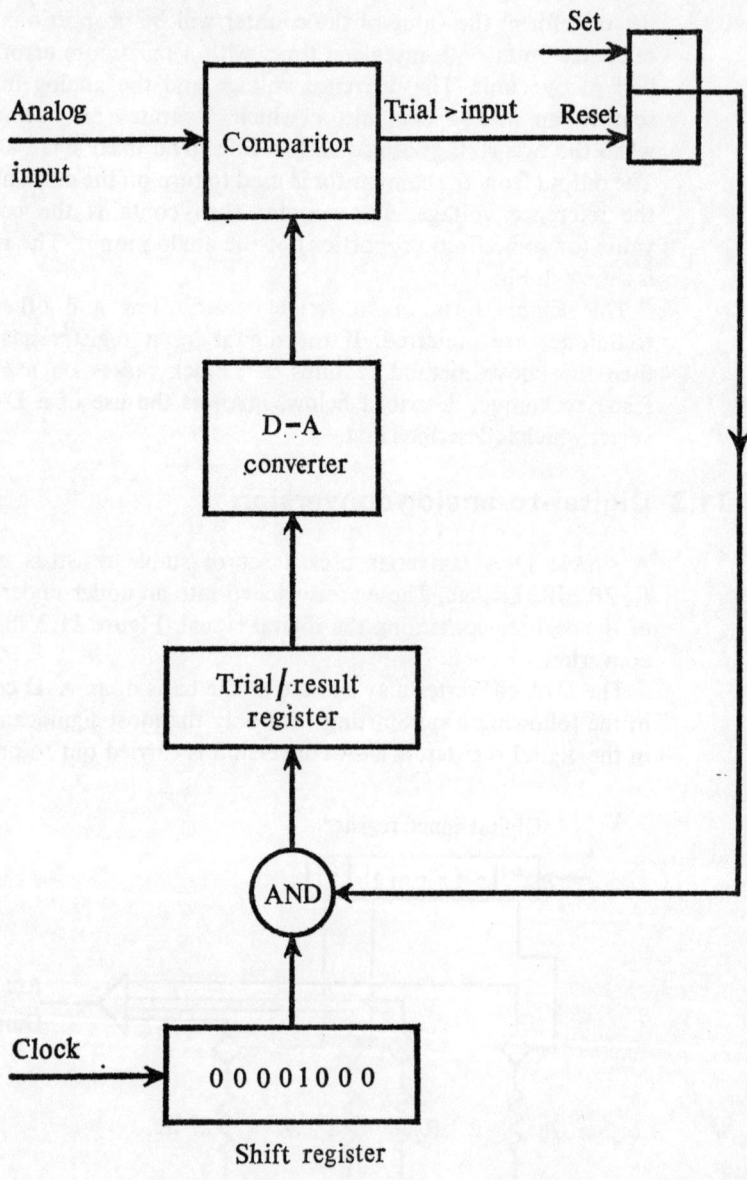

Fig. 11.3

ANALOG AND PARALLEL INTERFACES

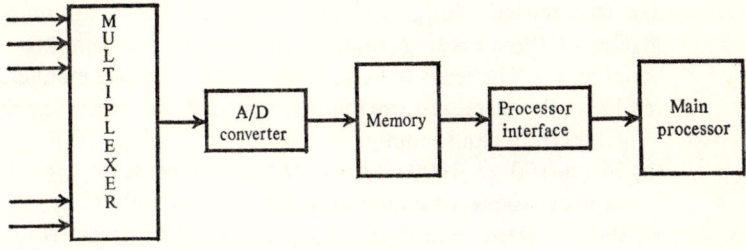

Fig. 11.4

trial voltage equal (proportional) to the value in the register. This trial voltage is compared with the input voltage. If the trial voltage is greater than the input voltage, the comparitor generates a pulse which is used to reset the current bit in the register to zero. Following the most significant bit, the next most significant bit is treated in the same way and so on down to the least significant bit. A clock is used to select the next bit. This technique uses only n clock ticks (as opposed to an average of 2^{n-1} for the first method) and is obviously significantly faster. The converter is shown in Fig. 11.3.

In dealing with a number of analog inputs one may multiplex them and make use of a single A–D converter. In this case the overall system is similar to that shown in Fig. 11.4. In this diagram the memory may be that of a communication processor.

11.4 Software

Usually in an on-line, real-time data acquisition and control application there are many tasks, and program segments within tasks are swapped in and out of core for economy of core store. There are a number of proprietary software systems designed to provide such overlay facilities together with a programming language designed for data acquisition work. One such system is IDAC-8 (see ref. 11.1) and its forerunner DATAK (see ref. 11.2). An example, using the simpler, less sophisticated DATAK system, is given below to demonstrate the type of problem encountered and the way of expressing the solution in such a language.

Suppose that an instrument package containing pressure, temperature and salinity sensing instruments is lowered into the sea. Data

DATA TRANSMISSION

are transmitted along a single conductor cable continuously in groups of three readings and is brought into the computer using a serial buffer. The sea is to be sampled in the following manner.

(a) From the surface to 100 metres, record at each metre the pressure, temperature and salinity.

(b) From 100 to 1000 metres, record temperature, pressure and salinity whenever temperature changes by 0·05°C or salinity by 0·02%. Also record the three every 100 metres from 100 to 1000 metres.

The required program in DATAK is

```
         BUFR:PRES, TEMP, SAL
         FORM:1, PRES, TEMP, SAL
         : OUTP(1,DCTP)
1        : IFGR PRES,144; GOTO 2
         : IFGR(PRES—@PRES),1; OUTP(1,DCTP)
         : GOTO 1
2        : IFLS (PRES—@PRES), 144;'
         : IFLS (TEMP—@TEMP),5;'
         : IFLS (SAL—@SAL), 2; GOTO 2
         : OUTP(1,DCTP); GOTO 2

         END
```

The above program assumes that the sensors have the following precision; that is, unity is equal to the following:

    1 unit of pressure = 1 metre
    1 unit of temperature = 0·01 °C
    1 unit of salinity = 0·01 %

The program says, in effect:

BUFR: PRES, TEMP, SAL	three variables associated with each set of data from the buffer.
FORM: 1, PRES, TEMP, SAL	output the three together.
: OUTP(1,DCTP)	output in form 1 to magnetic tape.

ANALOG AND PARALLEL INTERFACES

1	: IFGR PRES, 144; GOTO 2	if absolute change in pressure greater than 100 (144_8) go to 2, otherwise next line.
	: IFGR (PRES—@PRES),1;' OUTP(1,DCTP)	if the change on two consecutive readings is greater than 1 (metre) output values, otherwise next line and
	: GOTO 1	test again.
2	: IFLS (PRES—@PRES),144;' : IFLS (TEMP—@TEMP),5;' : IFLS (SAL—@SAL),2; GOTO 2	(the apostrophe means that the statement is continued on next line) if the pressure change is less than 100 or the temperature change is less than 0·05°C or salinity change is less than 0·02% test agian
	: OUTP(1,DCTP); GOTO 2	otherwise output and then test again.
	END	end of program!

Obviously such a language makes the solution of data acquisition problems considerable easier than working in assembly language. The example given above is a simple one which does not exhibit the full power of the language and indeed DATAK has been superseded by a more powerful system giving the user a more flexible and convenient tool.

11.5 Parallel data transmission

The British Standard Interface (see ref. 11.3) is an example of a parallel device that imposes no timing restrictions upon either the transmitter or the receiver. It is illustrated in Fig. 11.5. The way in which each of the 18 lines is used is illustrated in Fig. 11.6.

DATA TRANSMISSION

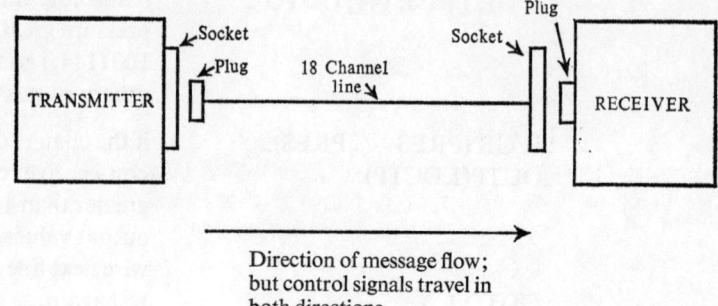

Fig. 11.5

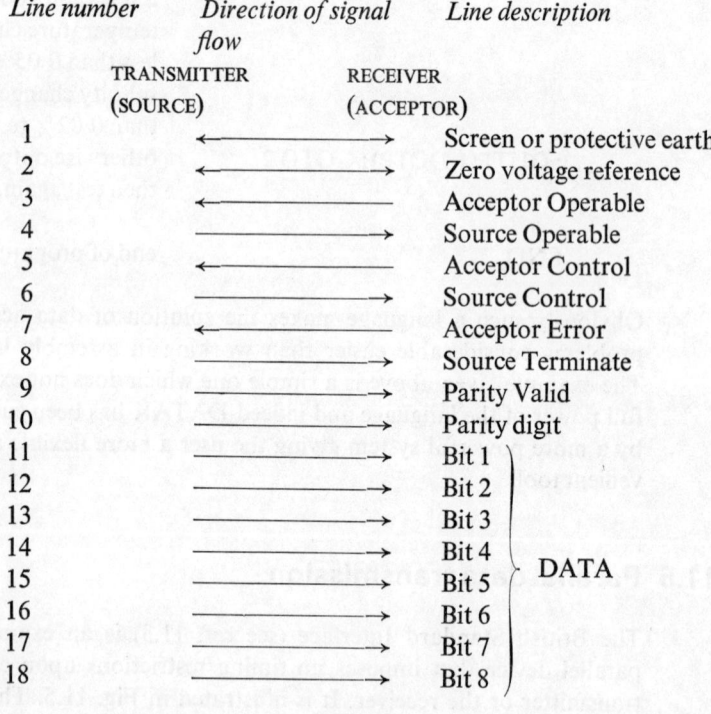

Fig. 11.6

ANALOG AND PARALLEL INTERFACES

The basic specification is for parallel transmission in the form of 8-bit characters from any transmitter to any receiver. Two interfaces are required for full duplex working. Line 1 joins the chassis of transmitter and receiver. Line 2 carries a potential to which all other lines are referred. The remaining lines carry binary signals. Lines 3 to 8 control the transmission of characters.

Lines 9 and 10: Line 9 indicates that parity is being used, line 10 being the parity bit. The transmitter can thus enable or disable the parity circuits in the receiver, and the stream of characters can be mixed, some with parity and some without, in any order. Lines 11 to 18 carry the information as an 8-bit character. Lines 3 and 4 are logic 1 when the device is usable and logic 0 when not usable: these are normally connected with the power on/off circuits.

Assuming that the transmitter and the receiver have been set, data is transmitted using a handshake technique, as shown in Fig. 11.7. The receiver requests the data by setting line 5 to logic 1; until this

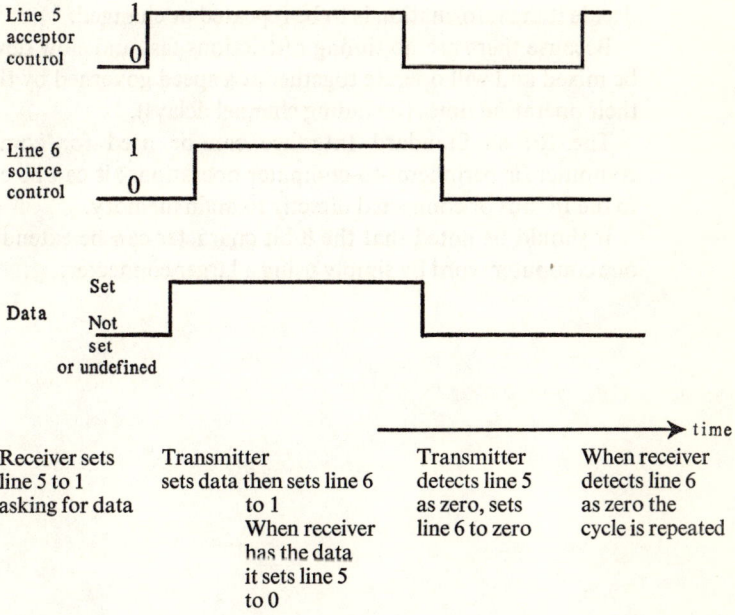

Fig. 11.7

occurs the transmitter must maintain line 6 at logic 0. When the transmitter detects logic 1 on line 5 it sets the data character; the transmitter can then set line 6 to logic 1, indicating that the message character may be read. The receiver, when it detects the 1 on line 6 can then read from the data lines, taking as long as it needs to do so There are therefore no timing restrictions. Once transmission commences, the transmitter must hold the data or message character steady for as long as a logic 1 remains on line 5. When the receiver has˙ finished reading the character it sets line 5 to logic 0, then and only then may the transmitter enter a new data character and change line 6 to logic 0. Line 7 can be used by the receiver to indicate an error (e.g. parity); it is usually used to request retransmission of the data. The receiver will not read data while line 6 is 0. When the receiver detects that line 6 is logic 0, the cycle may be started again. Once line 5 is 0, the transmitter can set a data character whenever it requires.

When line 7 is not used the transmitter can prepare new data as soon as line 5 changes to logic 0. When line 7 is used, the transmitter must wait until line 5 goes to 0, so that line 7 can be interrogated to decide if the information is to be repeated or changed.

Because there are no timing restrictions fast and slow devices may be mixed and will operate together at a speed governed by the sum of their operating times (including channel delays).

The British Standard Interface can be used for computer-to-computer or peripheral-to-computer operations; it can be connected to the I/0 bus or connected directly to main memory.

It should be noted that the 8-bit character can be extended to say one computer word by simply using a larger connecter.

12 Telecommunication organisations

12.1 International Telecommunication Union

The main international body concerned with telecommunications is the International Telecommunication Union (ITU) which is a specialised agency of the United Nations Organisation. The aims of the ITU are to: 'Maintain and extend international co-operation for the improvement and rational use of telecommunication; promote development of technical facilities and their most efficient operation. In order to improve efficiency of telecommunication services, increase their usefulness, and make them, as far as possible, generally available; harmonise the actions of nations in the attainment of the

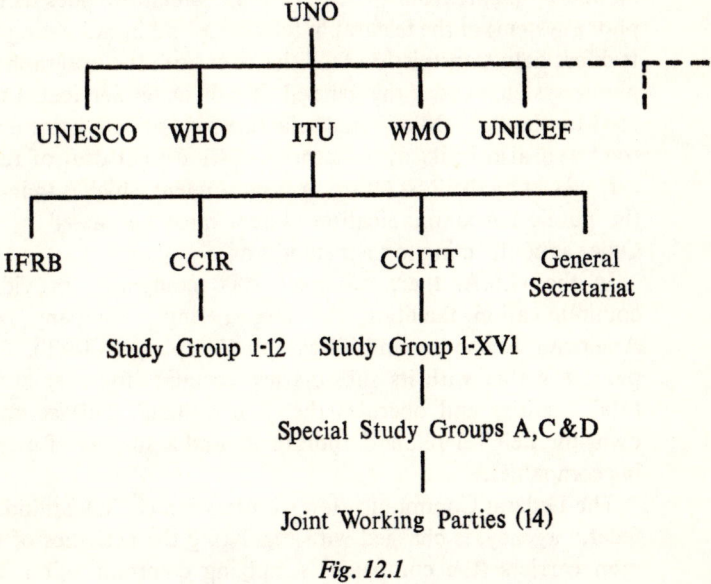

Fig. 12.1

DATA TRANSMISSION

common ends.' Membership is open to any country, territory or group of territories. Apart from the general secretariat, there are three permanent organs—International Frequency Registration Board, International Radio Consultative Committee and the International Telegraph and Telephone Consultative Committee (CCITT).

The CCITT was set up in 1957 and its aims are to study and issue recommendations on technical, operating and tariff questions connected with telegraphy, facsimile and telephony. There are some 40 study groups, working parties and committees within the CCITT framework. The most relevant of these is special study group A (Sp-A) concerned with data transmission. Fig. 12.1 shows the structure of ITU.

It is largely due to the work of the ITU that the various national telecommunication networks exhibit so many similarities.

12.2 National organisations

In the U.K. the telecommunication facilities are provided by the Post Office, a government-regulated commercial firm, which has a total monopoly (apart from the city of Hull, which provides its own telephone system) of the telegraph, telephone and broadcasting facilities. It offers data transmission facilities over both the telegraph and telephone systems under the general title of Datel Services. Equipment used by the Post Office is manufactured by private firms under contract and also in its own factories, also a great deal of research is carried out by the Post Office. Any equipment which is to interface to the public telecommunication system must be passed by the Post Office as conforming to their standards.

In the U.S.A. there are over 2000 companies providing telecommunication facilities, the largest single company being the American Telephone and Telegraph Company (AT&T). This company, together with its subsidiaries, accounts for over 80% of the total facilities and operates the 'Bell System'. This company also owns the Bell Telephone Laboratories and a number of manufacturing companies.

The Federal Communication Commission (FCC), an independent federal agency, is charged with regulating the activities of the common carriers (the companies supplying communication facilities).

TELECOMMUNICATION ORGANISATIONS

Common carriers must obtain approval from the FCC before offering a public service. The FCC is also charged with controlling interstate and foreign facilities. Each state has a commission dealing with communication facilities within that state.

In most other countries communication is the responsibility of the postal authority or a government department.

12.3 International services

Both International Telephone and Telegraph Corporation and Radio Corporation of America Communications offer an international telegraph and Telex service to over 100 countries.

International telecommunication links are also provided by means of communication satellites operated by the International Telecommunications Satellite Consortium (INTELSAT). Over 50 nations are members of INTELSAT and each shares the development cost of the common satellite system. In the U.K. the Post Office is the participating organisation in INTELSAT. The corresponding organisation in the U.S.A. is COMSAT (Communication Satellite Corporation). COMSAT is the largest partner in INTELSAT and is responsible for most of the operation and development work on satellites for the group.

12.4 Survey of facilities

This section gives a brief outline of the type of services available at the time of writing. It is not intended to make this a comprehensive list of Post Office or other common carrier services, since all such information is readily available from the suppliers.

It is to be noted that CCITT make various recommendations on such topics as line characteristics, transmission rates and band subdivision, etc. For example, R31 recommends that a voice channel be subdivided to carry 24 telegraph (50 baud) channels. This uses 24 frequency bands from 120 Hz to 3200 Hz each of 120 Hz.

The various common carriers offer data transmission facilities over the public telephone and telegraph system as well as over private leased lines. These facilities cover a range of transmission speeds and

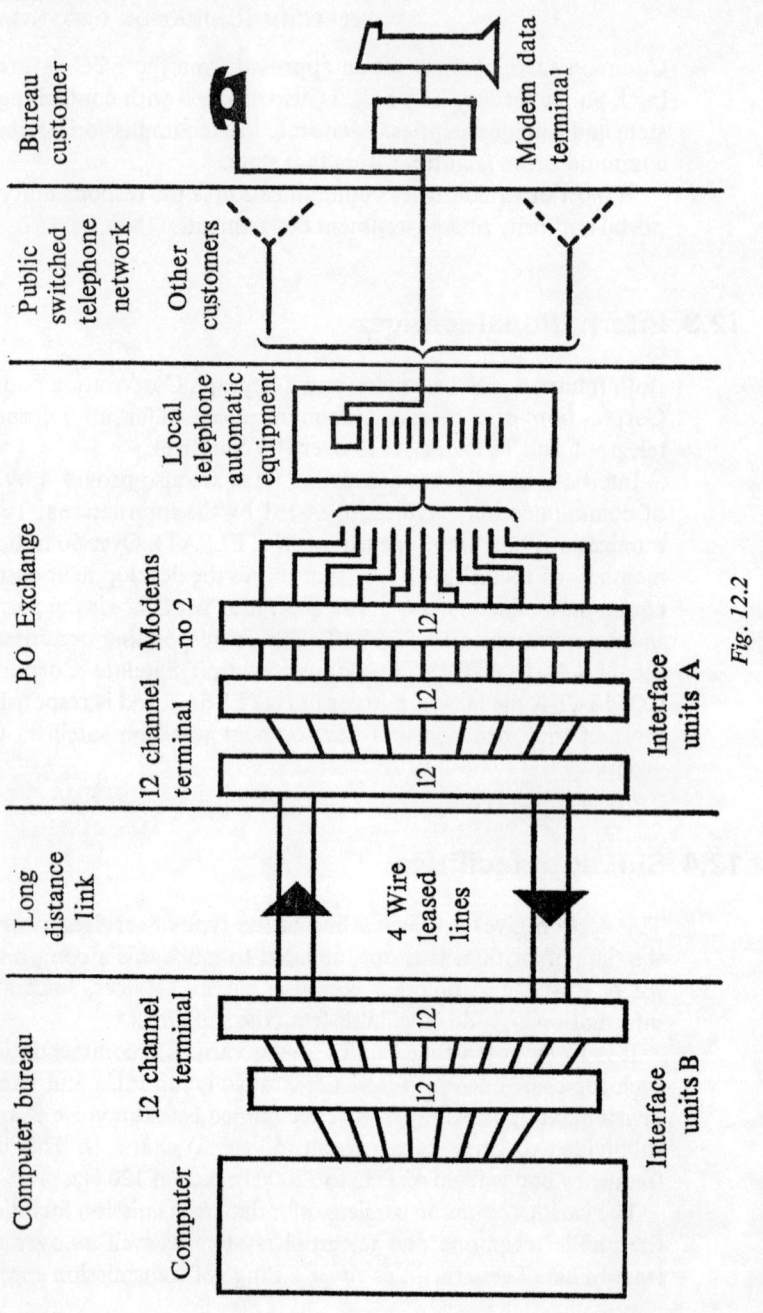

Fig. 12.2

TELECOMMUNICATION ORGANISATIONS

types of transmission. Most common carriers offer a range of modems. Some offer other services, for example, the British Post Office offers a 'Dataplex' service which provides for a group of terminals using the public switched telephone network to be concentrated and multiplexed onto a leased circuit to the computer centre; there they are expanded back to individual lines. Fig. 12.2 illustrates this.

The transmission lines offered by the common carriers may be broadly classified into the following categories.

(a) Narrow band lines

These are typically provided on telegraph-type lines. The speed range is from 45 to 300 bits per second. Thus, for example, the British Post Office offers lines at 50, and 200 bits per second (Datel 100 and 200). Most lines of this type are provided by means of a subdivided telephone channel, although none of the suppliers indicate the physical lines used, and indeed a given call may be routed over many different types of transmission lines.

(b) Voice-grade lines

These are provided on either public or private telephone circuits. The transmission rates are from 600 to 4800 bits per second. The public circuits are provided over the public switched telephone network. Currently, speeds in excess of 1200 bits per second require leased private lines which have been upgraded or conditioned in some way to improve the data transmission characteristics of the line (at an extra cost of course). In the U.S.A. lines of 600, 1200, 1400, 2400 and 4800 bits per second are available. In the U.K. lines of 600, 1200 (Datel 600) and 2400 (Datel 2400) bits per second are offered for data transmission.

(c) Wideband lines

These lines give speeds in the range 20–500 thousand bits per second. In the U.S.A. lines 19.2K, 40.8K, 105K, 240K and 500K bits per second are available, while in the U.K. the speed is 48K bits per second. Current research on optical waveguides promises transmission rates well in excess of these.

An interesting proposal has been put forward by the British Post Office for a new and separate digital data network. To provide for the

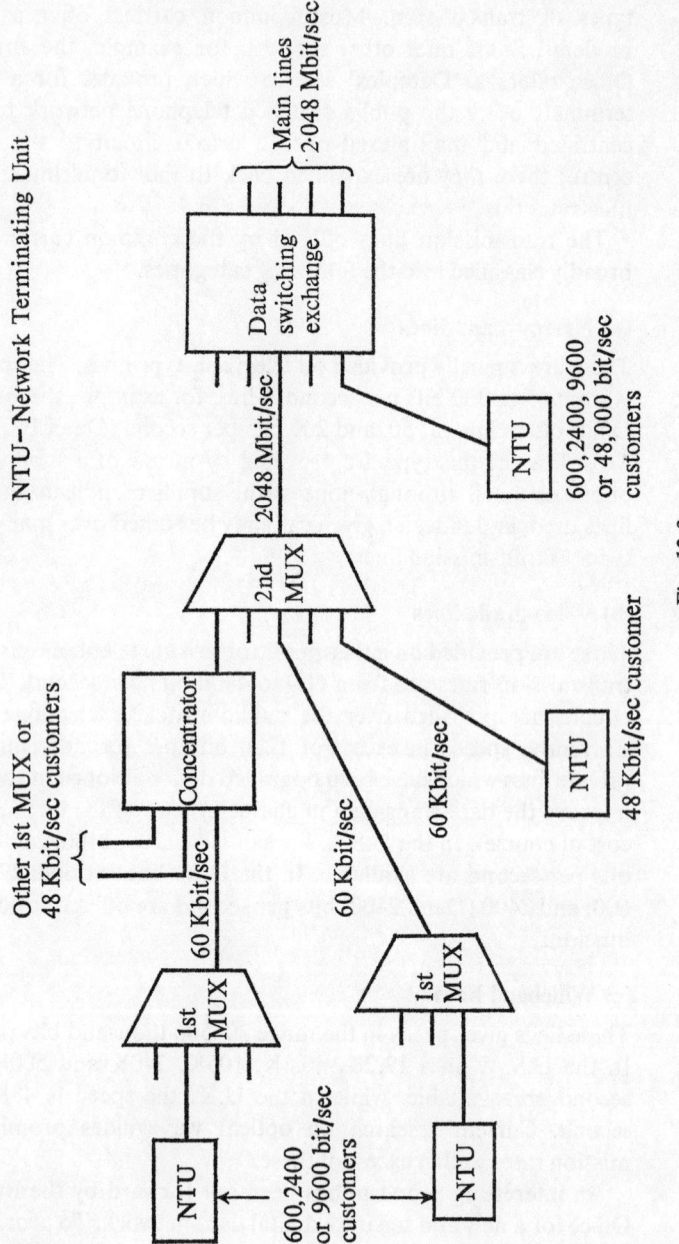

Fig. 12.3

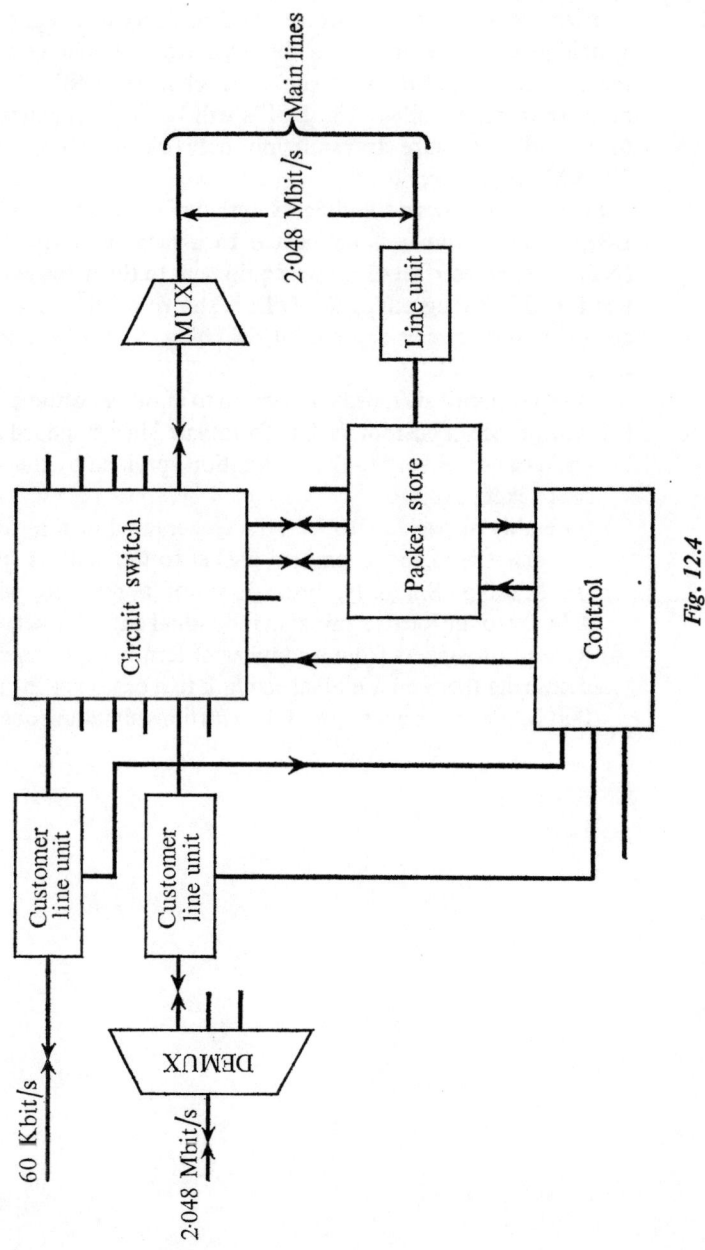

Fig. 12.4

DATA TRANSMISSION

interconnection of slow, medium and fast services, separate digital switching units are envisaged divorced from telephone exchanges; these would be called data switching exchanges (DSE's), and would be processor controlled. The DSE's will be liberally interconnected to provide a secure transmission network operating mainly at 2·048 Mbits per second.

Figure 12.3 shows a simplified diagram of the network feeding each DSE. Each customer is connected to a network terminating unit (NTU) which interfaces his own equipment to the network. A simplified functional diagram of the DSE is shown in Fig. 12.4. It will incorporate both circuit and packet switching. A DSE will provide the following facilities.

(a) A rapid circuit switched connection to another customer.
(b) Acceptance of customers data formulated in a standard addressed packet and delivery to the destination specified by the address in the packet heading.
(c) Assembly of packets from characters entered by a local terminal and delivery of the assembled packet to the address specified in the heading. Similarly, breakdown of a received packet and delivery to the local terminal in individual data characters.
(d) Receipt of packets from a number of terminals, interleaving and delivering them on a high-speed link to a customer in such a way that he may conduct several conversations simultaneously.

References

1.1 Lee, Y. W., *Introduction to the Statistical Theory of Communication* (Wiley 1964).
1.2 Brillouin, L., *Science and Information Theory* (Academic Press 1956).
1.3 Shannon, C. E., and Weaver, W., *The Mathematical Theory of Communication* (University of Illinois Press 1964).
1.4 Hartley, R. V. L., Theory of Information, *Bell System Technical Journal*, Vol. 7 (1928).
1.5 Nyquist, H., 'Certain Factors Affecting Telegraph Speed', *Transactions A.I.E.E.* (1924).
1.6 Nyquist, H., 'Topics on Telegraph Transmission Theory'. *Transactions A.I.E.E.* (1928).
1.7 Shannon, C. E., 'Mathematical Theory of Communication', *Bell System Technical Journal* (July 1948).

2.1 *American Standard Code for Information Interchange*. American Standards Association Inc. (June 1963).

3.1 Tucker, D. G., 'The Early History of Amplitude Modulation, Sidebands, and Frequency Division Multiplex', *The Radio and Electronic Engineer*, Vol. 41 (July 1971).
3.2 Tucker, D. G., 'The Invention of Frequency Modulation in 1902', *The Radio and Electronic Engineer*, Vol. 40 (July 1970).
3.3 (i) Dwight, H. B., *Tables of Integrals and other Mathematical Data* (Collier-Macmillan 1961).
    (ii) Bronwell, A., *Advanced Mathematics in Physics and Engineering* (McGraw-Hill 1953).
3.4 Lawson, J. L., and Uhlenbeck, G. E., The early chapters of *Threshold Signals* (McGraw-Hill, 1950).

DATA TRANSMISSION

4.1 Bass, C. A. (Ed.). 'Applications of Walsh Functions' *Proceedings of Symposium and Workshop, Naval Research Laboratory* (Washington D.C. April 1970).

4.2 'Application of Walsh Functions. Proceedings of Symposium and Workshop', Naval Research Lab. Washington. D.C. *Special Issue I.E.E.E. Transactions on Electromagnetic Compatibility*(August 1971).

4.3 'Theory and Applications of Walsh Functions', Proceedings of Symposium held at The Hatfield Polytechnic, Hatfield, Herts, U.K. June 1971.

5.1 Peterson, W. W., *Error Correcting Codes* (M.I.T. Press 1961).

5.2 Hamming, R. W., 'Error Detecting and Error Correcting Codes', *Bell System Technical Journal*, Vol. 29, No. 147 (1950).

5.3 Hagelbarger, D. W., 'Recurrent Codes: Easily Mechanised Burst Correcting Binary Codes', *Bell System Technical Journal*, Vol. 38 (1959).

5.4 Berlekamp, E. R., *Algebraic Coding Theory* (McGraw-Hill 1968).

5.5 Martin, J. W., *Telecommunications and the Computer* (Prentice-Hall 1969).

7.1 Barber, D. L. A., 'Experience with the use of the British Standard Interface in Computer Peripherals and Communication Systems'. *ACM Data Communications Symposium, Pine Mountain, Georgia* (October 1969).

7.2 Scantlebury, R. A., 'A Model for the Local Area of a Data Communication Network—Objectives and Hardware Organisation', *ACM Data Communications Symposium, Pine Mountain, Georgia* (October 1969).

7.3 Wilkinson, P. T., 'A Model for the Local Area of a Data Communication Network-Software Organisation', *ACM Data Communications Symposium, Pine Mountain, Georgia* (October, 1969).

7.4 Davies, D. W., 'The Principles of a Data Communication Network for Computers and Remote Peripherals', *Proceedings IFIP Symposium on Hardware* (Edinburgh 1968).

REFERENCES

8.1 Allen, S. G., 'A Comparison of p.c.m. and f.d.m.–f.m. Microwave Radio Systems', *The Radio and Electronic Engineer* (May 1971).

9.1 *DEC Communication Equipment*. Digital Equipment Corporation. 1971.

11.1 *IDAC-8 Programming Manual*. Digital Equipment Corporation 1969.

11.2 *DATAK Programming Manual*. Digital Equipment Corporation.

11.3 *A Digital Input/Output Interface for Data Collection Systems. BS 4421 :1969*. British Standards Institution.

Glossary of terms

This glossary is restricted to some of the important terms: it is intended to be used in conjunction with the index and the text. The major source of information is the CCITT *List of Definitions of Essential Telecommunications Terms*, published by the ITU, Geneva.

alphabet A table of corresponding terms relating an agreed set of characters and the signals which represent them.

a.m. Amplitude modulation.

analog transmission The transmitted signal is a continuous signal as opposed to a discrete signal. The human voice is an analog signal and can be transmitted as such, or can be coded to be a digital (discrete) signal.

audio frequencies Those frequencies to which the human ear is sensitive, approximately in the range 20 to 20,000 Hz.

ASCII American Standard Code for Information Interchange.

bandwidth The range of frequencies used.

baseband The naturally occurring range of frequencies; e.g., those actually produced by the human voice.

bit Contraction of the words binary digit.

buffer A storage device. Used to compensate for a difference in the rate of data flow.

carrier A single frequency capable of being modulated with an information signal.

character Letter, figure, number or other sign contained in a message.

code A system of symbols used to represent information.

common carrier A term used mainly in the U.S.A. for a Company which offers communications services to the public.

GLOSSARY OF TERMS

compandor A device which reduces the volume range of signals. It is usually followed by an **expandor** to restore the signal to its original form. The purpose is to improve the signal to noise ratio of the signal transmitted.

Dataphone A trade mark of the A. T. & T. Company. It is used to identify the data sets used in the transmission of data over the regular telephone network.

data set A modem.

demodulation The process of retrieving the information from a modulated carrier wave.

drop The line from a telephone cable to a subscriber's building.

error Any corruption of a message.

facsimile (fax) The transmission of pictures, maps, diagrams, etc.

f.d.m. Frequency division multiplex.

f.m. Frequency modulation.

F.C.C. Federal Communications Commission.

four wire circuit A transmitter and receiver are connected by two pairs of conducters, one for a 'go' channel, the other for a 'return' channel.

f.s.k. Frequency shift keying. A method of using frequency modulation to transmit digital data.

interface A shared boundary. For example, the boundary between two subsystems or devices.

isochronous transmission A bit string is Isochronous if the time intervals separating successive significant intervals are whole multiples of a unit interval. Thus Isochronous Transmission is a special case of Synchronous Transmission, the time intervals in the transmitted signal are always whole multiples of (for example) one bit. The timing/synchronisation is easier to implement.

ITU International Telecommunication Union.

KSR Keyboard Send/Receive.

LDX Long-distance Xerography.

line switching Switching together two lines temporarily, in order that signals may pass directly.

LRC Longitudinal Redundancy Check.

message switching The switching technique of receiving a message, storing it until the proper outgoing circuit is available and then retransmitting it to its destination.

modem Contraction of the words modulator/demodulator.

DATA TRANSMISSION

multidrop Line or circuit interconnecting several stations.
OCR Optical Character Recognition.
off line Not in the line loop. Not under control of the central processing unit. Terminal equipment not connected to a transmission line.
on line Directly in the line loop. Pertaining to devices in direct communication with the central processing unit. Terminal equipment connected to a transmission line.
PABX Private Automatic Branch Exchange.
PAX Private Automatic Exchange.
PBX Private Branch Exchange.
p.c.m. Pulse code modulation.
p.d.m. Pulse duration (or width) modulation.
p.m. Phase modulation.
p.p.m. Pulse position modulation.
private line (or wire) A channel or circuit provided for a subscriber's exclusive use.
p.w.m. Pulse width modulation.
signal Any intentional excitation of a circuit in a communication system.
store-and-forward The process of message handling in a message-switching system.
stunt box A device to handle the non-printing characters of a teleprinter, e.g. carriage return, line feed.
STD Subscriber Trunk Dialling.
tariff The contract between the customer and the common carrier. The published rate for a service provided by the common carrier.
telegraphy A communication system for transmitting graphic symbols (usually letters or numerals) by use of a signal code.
teleprinter The printing equipment in telegraphy. A teletypewriter.
Teletype Trademark of the Teletype Corporation. Usually refers to a series of different types of teleprinter equipment.
Telex A dial-up telegraph service. It operates world wide.
TEX Telex.
t.d.m. Time division multiplex.
time sharing A technique whereby a number of users share a computer facility at (apparently) the same time.
trunk (Britain) toll (USA) circuit A circuit connecting two exchanges in different localities.

TWX Teletypewriter Exchange Service.
vogad Voice operated gain adjusting device. A device similar to a compandor.
voice frequency Any frequency in the audio range essential for the transmission of speech. Usually in the range 300 to 3500 Hz approximately.
voice grade channel A channel suitable for the transmission of speech or other voice frequency signal.
WATS Wide Area Telephone Service.
wideband channel A channel wider in bandwidth than a voice grade channel.
word In telegraphy six characters plus one space. In computing an ordered set of characters that can be treated as the unit; it may be capable of being stored in one location.

Bibliography

Bell, C. G., and Newell, A. (Eds.), *Computer Structures: Readings and Examples* (McGraw-Hill 1971).

Bell System Technical Journal, Papers on Optical Waveguides (January 1971).

Berlekamp, E. R., *Algebraic Coding Theory* (McGraw-Hill 1968).

Brillouin, L., *Science and Information Theory* (Academic Press 1956).

Bull, G. M., and Packham, S. G. F., *Time Sharing Systems* (McGraw-Hill 1971).

Burroughs Corporation, *Digital Computer Principles* (1969).

Cherry, C., *On Human Communication* (Wiley 1957).

Digital Equipment Corporation, *DEC Communications Handbook* (1970).

Downing, J. J., *Modulation Systems and Noise* (Prentice Hall 1964).

Fano, R. M., *Transmission of Information* (MIT Press 1965).

Flores, I., *Computer Organisation* (Prentice Hall 1969).

Flores, I., *Computer Design* (Prentice Hall 1967).

Fraser, W., *Telecommunications* (Macdonald 1963).

Freebody, J. W., *Telegraphy* (Pitman 1959).

Gentleman, W. M., and Sande, G., 'Fast Fourier Transforms—for Fun and Profit', *Proceedings Fall Joint Computer Conference* (1966).

Lathi, B. P., *Communication Systems* (Wiley 1968).

Lee, Y. W., *Introduction to the Statistical Theory of Communication* (Wiley 1964).

Lucky, R. W., Salz, J., Weldon, E. J., *Principles of Data Communications* (McGraw-Hill 1968).

Martin, James, *Telecommunications and the Computer* (Prentice Hall 1969).

BIBLIOGRAPHY

Middleton, D., *Introduction to the Statistical Theory of Communication* (McGraw-Hill 1960).
Murphy, D. E., and Kallis S. A., *Introduction to Data Communication* (DEC 1968).
Panter, P. F., *Modulation Noise and Spectral Analysis* (McGraw-Hill 1965).
Peterson, W. W., *Error Correcting Codes* (MIT Press 1961).
Pierce, J. R., *Symbols, Signals and Noise* (Hutchinson 1962).
Scott, N. R., *Electronic Computer Technology* (McGraw-Hill 1970).
Shannon, C. E., and Weaver, W., *Mathematical Theory of Communication* (University of Illinois Press 1964).
Schwartz, M., *Information Transmission Modulation and Noise* (McGraw-Hill 1970).
Uhlenbeck, G. E., and Lawson, J., *Threshold Signals* (McGraw-Hill 1950).

Index

AM; *see* amplitude modulation
Amplitude, 12
Amplitude modulation, 26
Analog interface, 108
Analog-to-digital conversion, 108
Angular frequency, 3
Aperiodic, 9
ASCII, 18
Asynchronous line interface, 79
Asynchronous transmission, 59
AT & T, 118

Band, 10
Band-limit, 10
Band-width, 10, 27
Bans, 13
Base band, 27
Baud, 61
Baudot code, 18
BCD; *see* binary coded decimal
Bell System, 118
Bessel function, 32
Binary coded decimal, 20
Bit, 13
British Post Office, 118
British Standard Interface, 113

Carrier frequency, 27
Carrier wave, 27
CCITT, 118
Channel, 57
Channel capacity, 15
Character assembly, 95
Character detection, 79
Circuit, 57
Circuit switching, 65
Clock, 79
Coaxial cable, 17
Common carrier, 119
Common memory, 87
Communication codes, 18

COMSAT, 119
Concentrator, 75
Continuous spectrum, 9
Control characters, 22, 98
Cross talk, 46

Data transmission facilities, 119
Datel services, 121
Decibels, 16
Digital-to-analog conversion, 109
Distortion, 47

EBCDIC, 21
Error detection and correction, 49
Excess-three code, 21

Facility, 57
Fast circuit switching, 66
FM; *see* frequency modulatio
Fourier analysis, 5
Fourier integral, 9
Fourier transform, 9
Four-out-of-seven code, 54
Frequency, 3
Frequency analysis, 5
Frequency domain, 9
Frequency modulation, 31
Full duplex, 57

Gaussian noise, 45
Gray code, 21

Haar functions, 44
Half duplex, 57
Hand shake, 66, 99, 115
Harmonic analysis, 5

IDAC—8, 111
Information, 12
Impulsive noise, 46
Input/Output bus, 69

INDEX

INTELSAT, 119
Intermodulation distortion, 46
International Telecommunication
 Union (ITU), 117
ISO code, 19
IT & T (International Telephone &
 Telegraph Corporation), 119

Line, 57
Line sharing, 75
Line spectrum, 9
Line status, 88

Mark, 59
Message switching, 66
Modem, 62
Modulation, 2
Modulation index (factor), 27
Multidrop network, 77
Multiplexers, 69
Multiplexing, 68
Multitone transmission, 64

Noise, 45
Noiseless channel, 15
Noise power, 16

Orthogonal digital carriers, 44

P.a.m.; see pulse amplitude modulation
Packet switching, 66
Parallel transmission, 57
Parity checks, 51
Path, 57
P.c.m.; see pulse code modulation
PDP8, 88
Periodic, 2
Periodic function, 6, 9
Phase, 4
Phase modulation, 34
P.m.; see phase modulation
Polling message, 78
P.p.m.; see pulse position modulation
Pulse amplitude modulation, 37
Pulse code modulation, 41

Pulse position modulation, 40
Pulse techniques, 36
Pulse width modulation, 40
P.w.m.; see pulse width modulation

Quantisation noise, 43

Radio frequencies, 26
R.C.A. (Radio Corporation of
 America), 119
Redundancy, 14, 50

Sampling frequency, 37
Sampling interval, 37
Select message, 77
Serial transmission, 59
Shannon, 14, 15
Signal power, 16
Simplex, 57
Software, 95, 103
Space, 59
Spectrum, 9
Start element, 59
Stop element, 59
Store and forward, 66
Stunt characters, 18
Sub-bands, 64
Synchronisation, 59, 60, 99
Synchronous line interface, 98
Synchronous transmission, 60

Time domain, 9
Thermal noise, 45
Through circuit switching, 65
Transmission media, 17
Turn around, 98

Walsh functions, 44
Waveguides, 17
Wavelength, 2
Waves, 1
White noise, 45
Wire, 57
Wireless transmission, 25

135

TK
5102.5
B29

SEP 30 1974

~~JUN 12 1974~~